Sky Detective

Investigating the Mysteries of Space

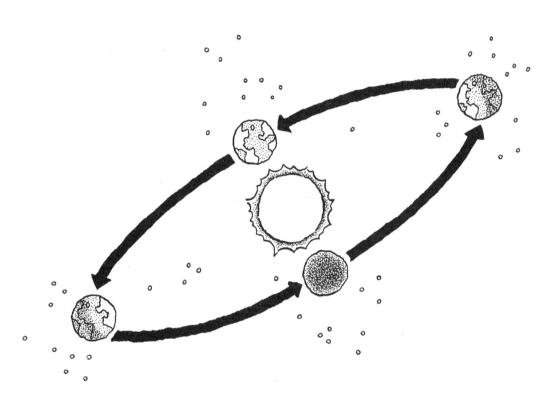

Eileen M. Docekal

Illustrated by David Eames

Sterling Publishing Co., Inc.   New York

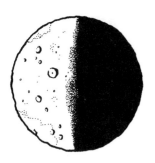

To my nieces and nephews: Jeremy
Jenna
Kerri
Kristin
Erica

Edited by Claire Bazinet

Library of Congress Cataloging-in-Publication Data

Docekal, Eileen M.
    Sky detective : investigating the mysteries of space / Eileen M.
Docekal ; illustrated by David Eames.
       p.   cm.
    Includes index.
    Summary: Text and suggested activities help the reader explore the
many aspects of the night sky, including the stars, constellations,
and planets.
    ISBN 0-8069-8404-X
    1. Astronomy—Observers' manuals—Juvenile literature.
2. Astronomy—Juvenile literature. [1. Astronomy.] I. Eames,
David, 1965–   ill. II. Title.
QB63.D55   1992
520′.22′2—dc20
                                                         91-46337
                                                         CIP

10 9 8 7 6 5 4 3 2 1   AC

First paperback edition published in 1993 by
Sterling Publishing Company, Inc.
387 Park Avenue South, New York, N.Y. 10016
© 1992 by Eileen M. Docekal
Distributed in Canada by Sterling Publishing
% Canadian Manda Group, P.O. Box 920, Station U
Toronto, Ontario, Canada M8Z 5P9
Distributed in Great Britain and Europe by Cassell PLC
Villiers House, 41/47 Strand, London WC2N 5JE, England
Distributed in Australia by Capricorn Link Ltd.
P.O. Box 665, Lane Cove, NSW 2066
Manufactured in the United States of America

Sterling ISBN 0-8069-8404-X Trade
              0-8069-8405-8 Paper

CONTENTS

INTRODUCTION

"Why do stars come out at night? What are fireballs, red giants, and white dwarfs? Why did the Pawnee Indians arrange their campfires according to the star patterns they saw in the sky? Why does the moon sometimes disappear?"

You can unravel these and other secrets of the stars, the moon, galaxies, and planets by becoming a sky detective, with the help of this book. Sky detectives hunt for treasures in the night sky. Your eyes, a clear night, and a bit of imagination are all you need to begin your investigations. The mysteries of the night sky can be solved anywhere, no matter if you live in the city or in the country. You can see the stars best from the top of a hill or an open field or park away from city lights. Street lamps, pollutants, and smoke often make stargazing difficult in the city.

Some of these sky probes can only be done at certain times of the year; other mysteries can be solved on any clear night. The investigations can be done from anywhere on Earth. If you live in England, the United States, or Canada, however, your view of the night sky will be different from what someone living in Australia, New Zealand, or South Africa sees.

Detecting Equipment

MOST IMPORTANT TOOLS

Your eyes   The most wonderful instruments for observing the sky.
Sighting tube   To help you see some of the dimmer stars when there are bright lights nearby. The cardboard tube from inside a roll of paper towels or gift wrap makes a good sighting tube.
Warm clothes   Bundle up for going out stargazing. Even in the summertime, nights can be cool so be sure to dress warmly. Think about taking along a hat, scarf, gloves, sweater, and thick socks.
Reclining lawn chair and blanket, or sleeping bag   Get comfortable and cozy when stargazing. It's hard to stand and look up for very long.

Flashlight and this book Place a paper bag or some red cellophane over the head of the flashlight. You will be able to see well enough to read, but won't have to wait so long for your eyes to adjust from the light to the dark.

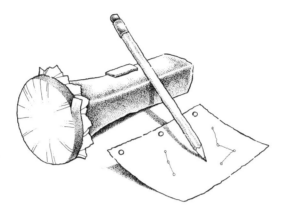

Sky detecting equipment

OTHER TOOLS

Binoculars These will help you find some of the dimmer heavenly sights. Support the binoculars by resting your elbows on the arms of a reclining chair, or a fence or large rock, or by mounting the binoculars on a tripod. The stars will "dance" when you are trying to see them if you just hold the binoculars up in front of your eyes.

> **Warning:** Never look directly at the sun or you can damage your eyes badly. If you look at the sun with binoculars or through a camera lens, you can go blind.

Notebook To write down unusual sightings with their date and time.
Camera You can photograph star trails, meteors—maybe, even a UFO!
Planisphere or star dial This handy chart will help you find out when constellations will be visible in your area. You will be able to "dial the sky" for any hour of the year.

•1•
A TREASURE CHEST OF DIAMONDS

Long ago, people looked up at the heavens with awe and wondered. Where did the sun disappear to each night? What were those strange objects that sparkled like diamonds in the darkness? Where did they come from? To explain what they saw, some of the people made up stories that were told and retold as the tribes gathered around their campfires, tended their sheep, or sailed the oceans—and the legends were born.

The early Egyptians believed that the sun was the great god Ra, who travelled in a giant boat across the wide blue river of the sky every day. To the Tsimshians, a Pacific Northwest tribe of Native Americans, the sun is The-One-Who-Walks-Over-the-Sky. He wears a flaming mask that lights the Earth. The stars form from the sparks that blow out of his mouth while he sleeps.

Tsimshian Sun Mask

Today, modern telescopes and the beginnings of space exploration have solved some of the riddles about the stars and planets. But the night sky is still full of secrets, and legends continue to be told. As a sky detective, you can probe the mystery and the myth to uncover clues about the workings of the universe. Just step out into the dark night and turn your eyes towards the heavens.

The Night Sky

Before you begin your search of the night sky, find a spot away from bright lights and get comfortable. Lie back in a lawn chair or down on a blanket and prop a pillow or knapsack under your head. Then close your eyes for a few minutes. When you open them, your eyes will be "dark adapted" to better see the stars.

COUNT THE STARS

Look up at the heavens and try counting the stars. Do you see a hundred, five hundred, or thousands of stars? There are so many, it's hard to keep track. Well, in case anyone should ask, on a dark, clear night, away from city lights, you can see about three thousand stars with your eyes alone. With binoculars, you can see about thirty thousand stars; and with a small telescope, almost two million stars!

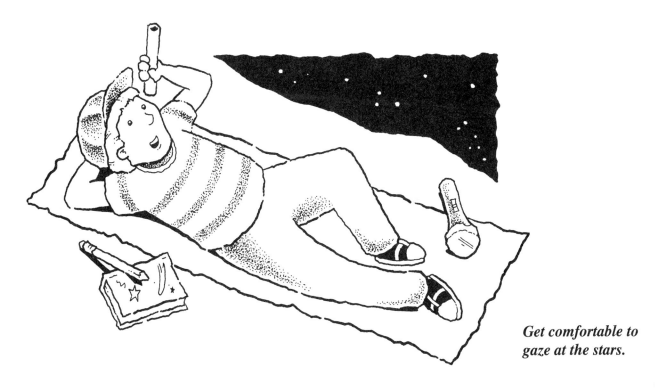

Get comfortable to gaze at the stars.

9

The Bushmen of Africa are a nomadic tribe that roam around the Kalahari Desert. There are no electric lights, or television, so they are very aware of the stars and moon that shine so brilliantly in the night sky. They believe that the stars are great hunters and have great power. When a son is born to a Bushmen mother, she asks the stars to accept the heart of her baby boy and give him some of their own heart instead so that he may be a brave and great hunter.

Bushmen mother raising baby to the stars

WHAT IS IT YOU SEE?

Have you ever wondered why the stars only "come out" at night? The stars are in the sky all the time but you can't see them during the day because they are hidden by the sun's brightness. Our sun is a star too, but it is much closer to us. The sun and all the other stars are giant balls of very hot gas, mostly helium and hydrogen. The starlight that you see comes from a series of nuclear reactions deep in the stars, something like what happens when a hydrogen bomb explodes. Tremendous amounts of energy are given off by the stars. The heat and the light that comes from just one star, the sun, provide the essential energy for life on Earth.

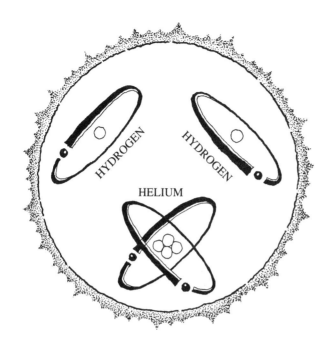

Sun with reactions. Starshine is light given off when hydrogen atoms combine to form helium.

MANY SIZES OF STARS

You may think of our sun as a gigantic object, but it isn't very large compared to many other stars. The sun looks large because it is close, only 93 million miles (150 million kilometres) away! Stars come in all sizes. The biggest stars are called supergiants and can measure almost 500 million miles (800 million kilometres)

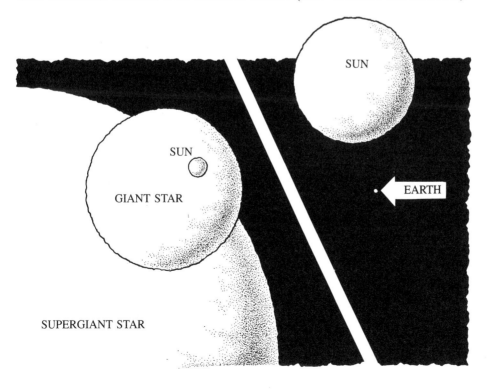

Star sizes. Our sun is large compared to the Earth (right) but small compared to giants and supergiants (left), which can be hundreds of times the size of our sun.

across. That's about the distance from the sun to the planet Jupiter. Neutron stars are the tiniest. They are only a few miles in diameter. Even though stars vary in size so much, they all look like tiny pinpoints of light to us because they are so very far from the Earth.

In fact, the distances to other stars is so great that *astronomers*, scientists who study the heavens, don't use miles or kilometres to measure them. Instead, they use *light-years*, the distance light travels in one year. Light travels very fast, about 186 thousand miles (3 million kilometres) every second. In a year, light travels close to 6 trillion miles (10 trillion kilometres).

A KIND OF TIME TRAVEL

When you look into the night sky do you realize you're looking back into time? It takes only eight minutes for the light from the sun to travel to the Earth, but the other stars are so far away that their light takes many years to reach us. For instance, the light rays that you see coming from the star Antares, a bright star found in the Northern Hemisphere in the summer sky and in the Southern Hemisphere in the winter sky, left that star 400 years ago, during the time Shakespeare was writing his plays. It is 400 light-years away from the Earth. Most of the starlight that you see has been travelling through space for centuries.

Starlight comes from long ago and far away.

DO THE STARS MOVE?

Find a group of stars and fix their position above a tall, stationary landmark, such as a telephone pole, roof, or treetop. Check on the position of this group of stars at different times during the night. Are the stars in the same place? You'll notice that that group and all the stars seem to move across the sky, rising in the east and sinking in the west as the evening goes on.

According to one Native American legend, the stars move because they are restless animals. They roam all over the sky hunting for better grasses and good weather.

Earth's spin makes stars seem to move across the night sky.

The movement of the sun, moon, and stars through the heavens was one of the greatest mysteries of all time. For many centuries, the thinkers of the Egyptian, Babylonian, Roman, and Greek cultures tried to explain the movement of the heavenly bodies. Finally, a Polish astronomer, Nicholas Copernicus, proclaimed that it wasn't the stars but the Earth that was moving. This *rotation*, or turning of the earth on its axis, takes twenty-four hours, or one day. It is the Earth's rotation that causes night and day and makes it look as if the stars are moving far overhead. When the side of Earth where you are is turned towards the sun, you experience daytime. When it is turned away from the bright sun it is night, and you can see a treasure chest of stars in the dark night sky.

Day and Night

To solve the mystery of why you experience day and night, you'll need:

- an orange
- a couple of toothpicks
- a crayon
- a flashlight

Pretend that the orange is the Earth and the flashlight is the sun.

① Stick one toothpick partway into one side of the orange to represent the North Pole.

② Stick the other toothpick into the opposite end to represent the South Pole.

③ With the crayon, mark an X on the orange to show where you live on the surface of the "planet" (illus. **A**).

④ Light the flashlight, place it on a table, and hold the orange by the toothpicks so that the beam of light falls on its surface.

⑤ Rolling the toothpicks between your fingers, slowly spin the orange. Watch how your "home" moves from light into darkness (illus. **B** and **C**).

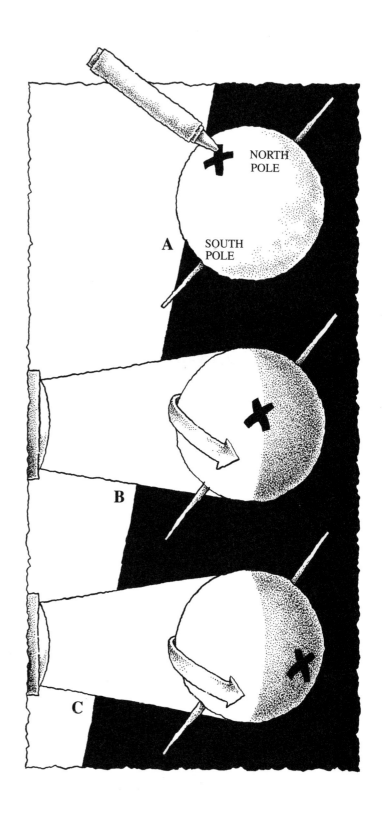

The Earth's Journey

The Earth does more than spin. It also travels in an oval or egg-shaped path around the sun. This journey, known as a *revolution*, takes 365-1/4 days, or one solar year, to complete. As the earth revolves, its *axis* is slightly slanted towards the sun. The axis is an imaginary line that connects the South Pole to the North Pole. It is the tilt of the axis that causes the length of the days to change from season to season during Earth's journey around the sun. When the

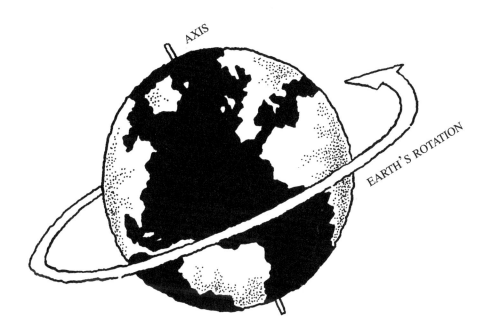

The Earth turns around a tilted axis.

North Pole is tilted toward the sun, the Northern Hemisphere experiences the long days of summer and the Southern Hemisphere has shorter winter days. When the North Pole is tilted away from the sun, it's winter in North America and Europe and summer in Australia and New Zealand.

The Pahute Indians of the Southwestern United States had a simpler explanation for the seasonal changes in the length of days. Sun Man is a spirit that carries the sun across the sky every day during the summer. He is old and feeble and his steps are short and slow. That's why summer days are long. During the cold days of winter, his young son takes up the sun, hurrying across the sky with the speed of youth. That's why winter days are short. As we

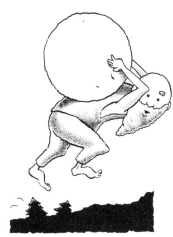

Pahute's Sun Man

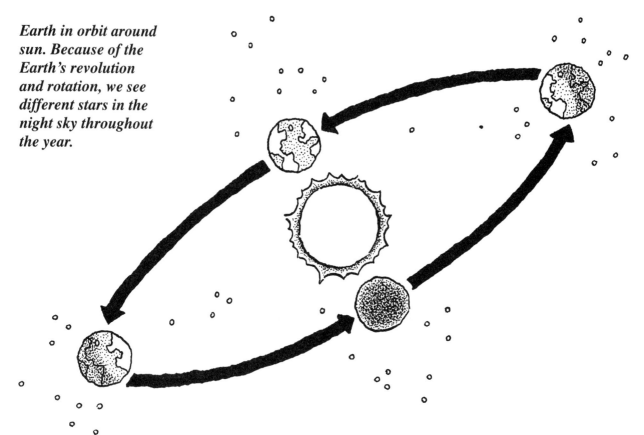

Earth in orbit around sun. Because of the Earth's revolution and rotation, we see different stars in the night sky throughout the year.

travel with the Earth around the sun, we move through different areas of the night sky. Because of the Earth's rotation and revolution, the stars we see, when we look up, change from hour to hour and from season to season through the course of the year.

Case #1: Moving Evidence

For solid proof that the stars exist, capture some of their twinkling light on film. You'll need:
- a camera that allows for time exposures
- a sturdy support for the camera. A tripod is best but a cement block, fence post, or rock will do.
- a cable release to hold the camera's shutter open
- fast film: either color or black and white, but preferably a film speed of ISO 400 or greater

To capture starlight on film:

① Position the camera on the support with the lens pointed sky-wards.

② Open the lens up to the widest f/stop (for example, f/2.8 or f/4.5) and set the focus to ∞ infinity.

③ Attach the cable release and set the shutter speed to "B" or "T" for a time exposure.

④ Remove the lens cap, then press the cable release and lock it in the open position.

⑤ Leave the camera open for about 5 minutes. Take care not to bump the camera.

⑥ Release the shutter and advance the film.

⑦ Shoot the next frames at different intervals; try 10 minutes, then 15 minutes, 20 minutes, and so on.

⑧ Record the frame number and how long the shutter was left open. This will help you figure out what times give the best results.

To capture star trails on film:

This activity will give you proof that you live on a world that is slowly spinning.

① Find a safe location to set up your camera.

② Press the cable release and leave the lens open for five to six hours. You may want to set an alarm and go to bed. Be sure to time the exposure so that you close the shutter before the light of dawn appears on the horizon.

You'll have the best results if you gather your photographic evidence in a rural area on a clear night without a moon. Also, when you drop your film off to be developed, tell the clerk or mark it on the envelope that the film is a special roll of star photographs. This will eliminate some confusion when they develop and print the dark negatives.

Star trails photographed from a location on the equator (left) *and from the poles* (right)

•2•
STAR-TRACKING

At first, the star-filled heavens may appear to be just a jumble of faraway lights. However, as you watch the sky from night to night, you will begin to recognize shapes and patterns among the stars. Some of these star patterns appear only during certain seasons. But you will see several of them in the heavens all year long. You can use them as guides to help you track your way around the night sky.

Discovering the Constellations

While lying on your back, fix your eyes on an area of the starry sky. It's easy to create pictures if you use your imagination to draw lines connecting the stars. These pictures, or star patterns, are called *constellations*. It's interesting to sketch some of the patterns you see in the sky.

MAKING A STAR FRAME

It's easier to sketch patterns if you use a star frame. You can make one out of a wire coat hanger.

Take a wire hanger and stretch it into a square by pulling on the bottom part of the hanger near the middle. Then make a handle by pressing on the hook to flatten it.

Cover the handle with masking tape so that you won't scratch yourself on the sharp wire ends.

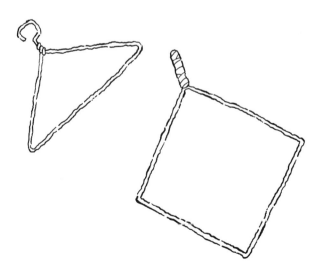

Now, grab a sheet of paper and a pencil, hold the frame up to the sky, and copy down the positions of the brightest stars you see inside the frame.

I SPY

What shapes do you see? Perhaps you can see a skateboard, cat, or a cartoon character floating in your night sky. People long ago usually recognized shapes of objects familiar to them. Hunters imagined buffalo and wolves, farmers saw ploughs and horses, and sailors identified compasses and sterns of ships.

Star patterns don't change. Once you become familiar with a constellation, whether it's one in this book or one you've created yourself, it will be a landmark for you. This makes sky detecting a lot easier. It's true that the constellations shift their positions throughout the night and the year. And many appear only during certain times of the year. But if you look into the night sky exactly one year from now, you'll see the same constellations in the same positions.

SKY CALENDAR

Our ancestors knew about the movement of the constellations and were able to use the night sky as a huge calendar. Farmers waited until certain constellations appeared in the sky before they planted seeds or harvested the wheat. The stars told hunters when to take up their bows and arrows and go in search of the migrating herds of antelope and buffalo to hunt for food. When certain stars rose overhead, the nomadic tribes of Africa journeyed to their annual gathering places, and Native Americans held their sacred ceremonial dances.

Travelers long ago used the stars as a guide.

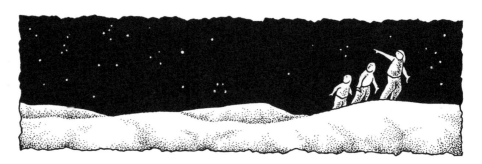

22

STORIES IN THE SKY

Imagine being an early explorer, travelling through the wilderness to strange new places and spending each night under a canopy of stars. During your travels, you discover the "skateboard" constellation. How would you remember your constellation so you can describe it and share your discovery with others?

Discovering the "skateboard" constellation

With a stick you could draw a map in the dirt, showing its place among the other stars. You could make up a story about how the skateboard got in the sky. The story would help you to remember the shape and be easy to pass along to your friends. That's how the ancient peoples told each other what they saw in the heavens. To them, the night sky was a storybook about ghosts and spirits from other worlds. Their tales were a way to make sense out of the mysterious night sky.

THE ORDER OF STARS

For the Pawnee Indians, who lived on the vast North American plains, the constellations were gods who guided many aspects of their lives. They believed that the stars were placed in their positions in the sky by the Heavenly Chief to protect the people who lived on Earth below. Like most Native American cultures, much of the life and religion of the Pawnee Indians emphasized living in harmony with nature. As a sign of respect for the order of stars and the celestial powers, the Pawnees arranged their campfires in the patterns of the constellations they saw in the night sky.

Pawnee campfires followed constellation patterns.

The North Star

If you live in the Northern Hemisphere, you may notice one star that doesn't move across the sky. This fixed star is the **North Star**, or *Polaris*. Polaris is not a very bright star but it is important because it sits directly over the North Pole. Sailors depended on Polaris to find their way at sea. If they saw the North Star moving higher in the sky each night during the course of their journey,

Sailor navigating by the stars

they knew they were travelling north. If it was moving down towards the horizon, they were heading south. There is no distinct star that sits over the South Pole, although many a sea captain who sailed the stormy southern oceans wished there was!

From China to Sweden, from North Africa to India, people looked to the North Star with awe and reverence. For them, the North Star was a jeweled spike that held the universe together. Without it, the stars, the moon, and the Earth would be flung off into the far reaches of outer space. The Pawnee Indians called the North Star simply The-Star-That-Does-Not-Move. It was a very sacred star, the chief spirit that watched over all the Indian tribes that lived on Earth and in the heavens.

Tracking Polaris

The best way to track down Polaris is to first find the unmistakable group of seven stars that forms the Big Dipper. Then, to find the pole star, trace a line from the two end stars in the bowl of the Big Dipper to a star that forms the handle of a smaller constellation, the Little Dipper. This is the North Star.

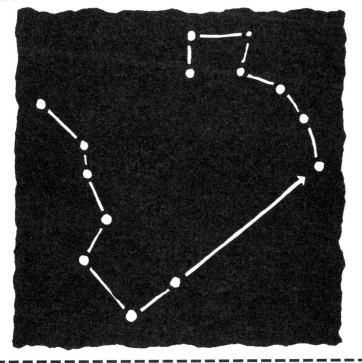

How far above your horizon is the North Star? If you live very far north, near the Arctic Circle, Polaris will be almost directly overhead. If you live along the equator, the North Star will be just above the horizon; and if you live anyplace south of the equator, Polaris will never be in your sky.

HUNTING THE BEAR

Both the ancient Greeks and Native Americans saw the stars in the Big Dipper as part of the Great Bear, *Ursa Major*. Ursa Major is the best known of all the constellations. It's easy to recognize the shape of the Big Dipper, but you'll probably have to use plenty of imagination to see the Great Bear.

Ursa Major. The Big Dipper is a part of the Great Bear. Look at the bear's unusually long tail to get a "handle" on it.

The North American Indians pictured a bear with three hunters stalking behind it. The first hunter carries a bow and arrow ready to take aim, the second carries a pot in which to cook the meat, and the third follows slowly gathering stars along the way so he can build a fire under the pot. They begin stalking the bear in the spring and finally shoot him in the autumn. The dripping of the bear's blood from the arrow's wound is what paints the leaves red in the fall. Because the bear is a spirit his soul never dies. Instead it reappears in the spring in the body of another bear and the hunt begins again.

SEEING STARS

How many stars can you see in the handle of the Big Dipper? The middle star in the Dipper's handle is actually a double star, two stars that appear to be very close to each other. These two separate stars can usually be clearly seen with binoculars. Arabs used this double star as a test of vision. The more visible star is **Mizar**, the horse, and next to it is its fainter partner, **Alcor**, the rider. If you could see the separate stars Alcor and Mizar, your vision was considered good.

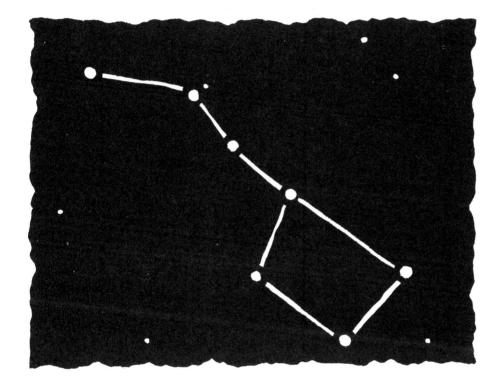

Big Dipper, with double stars Mizar and Alcor. Can you find them in the Big Dipper's handle?

In 1804, the famous British astronomer William Herschel figured out that the two stars in most double-star sets actually revolve around each other. Though most stars seem like one dot when seen with the eye, single stars are a minority. Over fifty percent of the stars you can see with your naked eyes are double- or multiple-star sets.

STAR LIGHT, HOW BRIGHT?

Here is another challenge for your eyes. Stars vary in brightness. The brightness of a star as it is seen by the average naked eye is called *magnitude*. The higher the magnitude number, the fainter the star. A star with a magnitude of −1 or 0 is extremely bright, while the faintest stars that can be seen using only your eyes on a dark, moonless night are of magnitude 6. With binoculars or telescope, of course, you can see much fainter stars.

Use the Big Dipper's star guides to find the Little Dipper (the Big Dipper seems to empty into the Little Dipper). Polaris is only a medium-bright star with a magnitude of 2. Can you spy the faint, 6th-magnitude star just outside the bowl of the Little Dipper?

Little Dipper showing star magnitudes. Try testing your eyesight: the stars that make up the Little Dipper vary in brightness from magnitude 2 to 6.

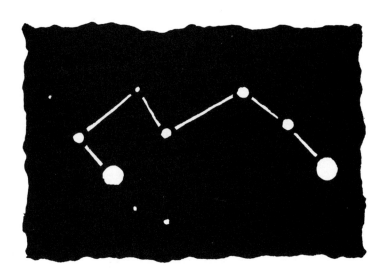

Northern Sky Constellations

During your evenings of stargazing, most stars will appear to rise in the east, trace an arc across the sky, and then set below the western horizon. However, you may notice (if you live north of Cairo or Miami) that the Big Dipper is there in your night sky at all times of the night, all year round. Unlike most constellations, the stars in the Big Dipper never disappear below the horizon. Instead, they move in a nightly circle around the North Star. These star patterns that are visible throughout the year are called *circumpolar constellations*. The Big Dipper is one of a handful of them found in the Northern Hemisphere. Circumpolar constellations are handy guides for exploring the rest of the heavens.

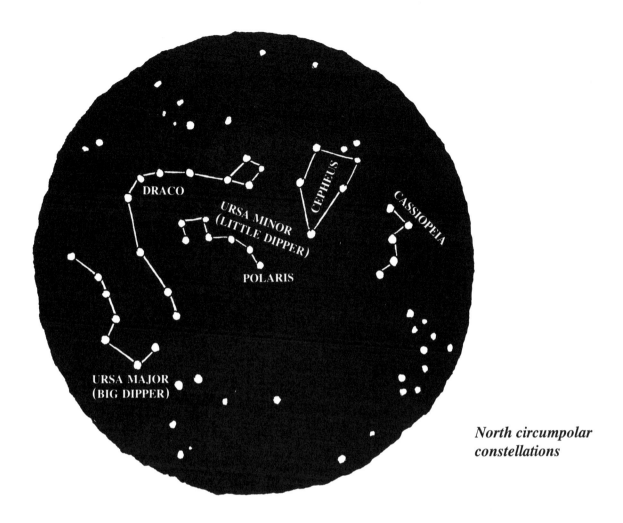

North circumpolar constellations

THE KING AND QUEEN

Cepheus, the King, is also a northern circumpolar constellation. The King's head is a rectangle and he seems to be wearing a dunce cap. Snaking between the Dippers and Cepheus is *Draco*, the Dragon. Close by is *Cassiopeia*, the wife of Cepheus. This constellation is shaped like a giant "W". Cassiopeia was the Queen of Ethiopia who boasted loudly of her beauty. This offended the sea nymphs, who appealed to the heavenly council to punish her. They voted to change her into stars that revolve in the sky. To shame her for her vanity, she must hang upside down on her throne six months of the year. Can you find Cassiopeia? How is she sitting on her throne? Cassiopeia sits high in the sky in the fall and is a good reference point for scouting autumn stars.

Cassiopeia on her throne

THE GREAT RA

Long ago, the circumpolar stars were believed to be where the spirits and gods resided in their palaces. The ancient Egyptians worshipped many gods, but the most important one was the great sun god, Ra. Ra travelled across the daytime sky in a magnificent boat. This huge ship required the labor of many strong oarsmen. The Egyptians believed the stars around the North Star were the mighty oarsmen of this boat, at rest in their sleeping quarters.

Ra, the Sun God, in his boat

Southern Sky Constellations

Do you live south of the equator? Then, the Southern Cross, or *Crux*, will be your chief circumpolar landmark. Unlike the North Pole, there are no brilliant stars near the South Pole. To find Crux, look for four bright stars arranged in a cross-like pattern. It actually looks more like a small kite or a diamond. When the earliest voyagers sailed around South Africa's Cape of Good Hope, this glowing cross was their most important star guide.

Almost surrounding Crux is an enormous constellation, *Centaurus*, named for a creature in Greek mythology that was half-man and half-beast. Two bright stars in this constellation, **Alpha** and **Beta Centauri**, point directly at the Southern Cross. Alpha Centauri is only 4.3 light-years away! It is the nearest star to us after the sun.

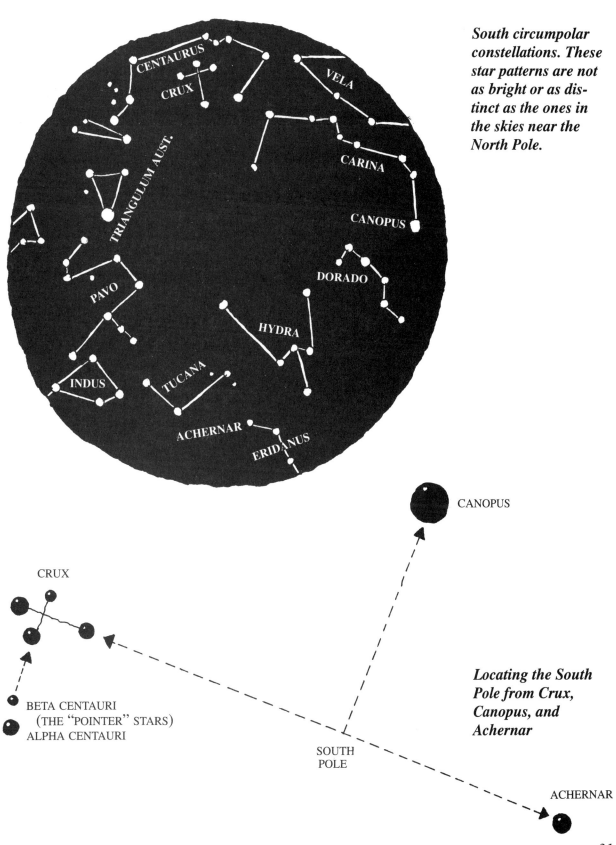

South circumpolar constellations. These star patterns are not as bright or as distinct as the ones in the skies near the North Pole.

CENTAURUS

CRUX

VELA

CARINA

CANOPUS

TRIANGULUM AUST.

DORADO

PAVO

HYDRA

INDUS

TUCANA

ACHERNAR

ERIDANUS

CANOPUS

CRUX

BETA CENTAURI
(THE "POINTER" STARS)
ALPHA CENTAURI

SOUTH
POLE

ACHERNAR

Locating the South Pole from Crux, Canopus, and Achernar

STAR BRIGHT

Can you find a south circumpolar star that seems to outshine its neighbors? This is the star **Canopus**, in the constellation *Carina*, the Ship's Keel. Canopus has a magnitude of −0.8, and is the second brightest star in the night sky. This dazzling star was widely worshipped in ancient Egypt. Many temples were built to face the spot on the horizon where Canopus rose along with the sun at the autumnal *equinox*, when day and night are equal. The Egyptians believed that it was Canopus that gave all precious stones their color and brilliance.

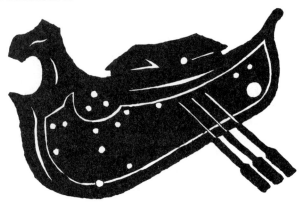

Carina, the Ship's Keel

As you probably noticed, many of the northern constellations and stars have Arabic or Greek names. Ptolemy, the famous astronomer in the courts of the last Egyptian Pharaohs, named 48 of the 88 constellations recognized in the sky today. The stars in the southern circumpolar region, which were mostly out of view of the Egyptians, Arabs, Greeks, and Romans, were mapped after the seventeenth century and have more modern names.

Case #2: Star Show

On rainy or overcast evenings you can bring the stars inside by making your own planetarium. You'll need:
- a large tin can or round cardboard salt or cereal box with both the top and bottom removed

- blank paper and pencil
- a straight pin
- scissors
- a rubber band
- a flashlight

To make the planetarium:

1. Use the bottom of the can or box to trace a few circles on paper.

2. Copy the circumpolar star patterns (see pages 29 and 31) in the circles, one constellation per circle (illus. **A**).

3. With the straight pin, carefully punch tiny holes where you marked the stars on the paper.

4. Cut around the circle, leaving about a 1/2 inch (1 cm) edge (illus. **B**).

5. Place the star pattern sheet paper over one end of the can and attach it with the rubber band.

6. Stick the flashlight inside the open end of the can or box, darken the room, and project your constellations onto a wall (illus. **C**).

This is a good way to get to know the constellations that will help you track the seasonal stars. You may want to give a star show for your family and friends. You can also tell or act out some of the stories about these constellations.

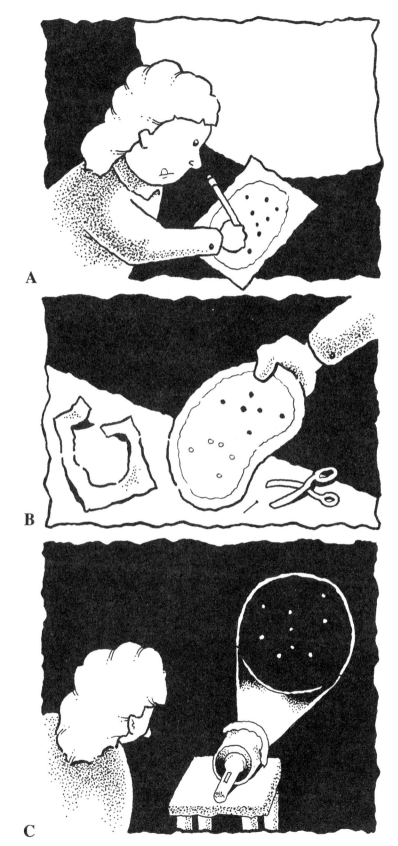

A

B

C

33

•3•
IN SEARCH OF THE
SKY SPIRITS

Most of the constellations in the night sky appear and disappear from view as the seasons change. Ancient stargazers and wizards charted these shifting patterns hoping to decipher the secret messages they contained. Some of them believed that the constellations were spirits that travelled through the heavens. The Greeks and Romans of long ago saw the heroes and heroines of their mythology, like Orion and Andromeda, or the ghosts of animals, like the swan and the scorpion. But these are only a few of the many ancient "spirits" you can discover hidden among the stars. Many of these night-sky residents contain some very special treasures, so be on the lookout for red giants, clusters, and stars that "wink."

The Winter Sky

It's best to begin your search for "the sky spirits" on a clear, moonless night. Winter is an excellent time to look at the stars because it gets dark earlier. Also, many winter constellations contain bright (first magnitude) stars that glitter through the cold, clear air. Even if you stargaze near strong city lighting, you can easily make out the major winter constellations. A cardboard tube makes an excellent spyglass for observing individual stars if you're close to street lights.

For those sky detectives who live south of the equator, the seasons listed for discovering the constellations are reversed. That is, the constellations described in this chapter as winter stars will appear in the summer night sky in the Southern Hemisphere, spring constellations will appear in the autumn sky, etc.

ORION, THE HUNTER

Whether you live in Alaska or Tasmania, you should have no trouble finding the most outstanding of all the constellations, *Orion*, the Hunter. Look for a large rectangle of four bright stars with three stars lined up near the middle. This distinctive trio of stars form his belt.

Orion, the Hunter

In this constellation, the Greeks saw the ghost of a fearless hunter marching across the sky with his club raised. According to one legend, the goddess Diana killed Orion by accident. Diana lived on the moon, but liked to travel down to the Earth to hunt. One day, while she was hunting, she saw the handsome hunter and fell in love with him. She refused to return to the moon, neglecting her job of keeping the heavens lit at night. This angered the sun god, Apollo, and he decided to trick her. Apollo challenged Diana to shoot at a dark figure far in the distance. Proud of her skill as a hunter, she aimed and shot an arrow that pierced the heart of the creature. When she came near and looked at her kill, she discovered it was Orion. Although she couldn't restore him to life, Diana made him immortal by placing him in the sky among the stars. There, she can see Orion as she drives her silver moon chariot across the night sky.

The moon goddess, Diana, kills Orion with an arrow.

Can you find the two brightest stars in Orion? One lies on his left shoulder, the other on his right ankle. With your naked eyes, you should be able to see a strong color contrast between them. The shoulder star is an orange-red star called **Betelgeuse**. The ankle star is **Rigel**, a brilliant blue-white star. (To those in the Southern Hemisphere, Orion appears to be standing on his head.)

STAR COLORS AND TEMPERATURE

Stars display a variety of colors so subtle that you don't notice them at first glance. If you look closely, you can see the range from white to blue, and from yellow to orange to a reddish shade. Star colors reveal something about the temperature of the gases in a star's outer layer. The hotter the star, the bluer it is. Blue-white stars are the hottest and red stars are the coolest. Rigel is a young, fast-burning star of a very hot temperature (about 35,000° F, 19,500° C).

As shown by its red color, Betelgeuse is a relatively cool star at only 5,000° F (2,800° C)! It is one of the largest stars known, so it is often referred to as one of the "red giants." If our sun were the size of Betelgeuse, it would stretch all the way to Mars and Earth would be inside it! Our sun is a medium-hot star, burning at about 11,000° F (6,000° C).

Betelgeuse, with four planets (including Earth) shown inside. A giant red star, Betelgeuse is thought to be about 500 times the size of our sun. If it were our sun, the orbits of Mercury, Venus, Earth, and Mars, would fit inside.

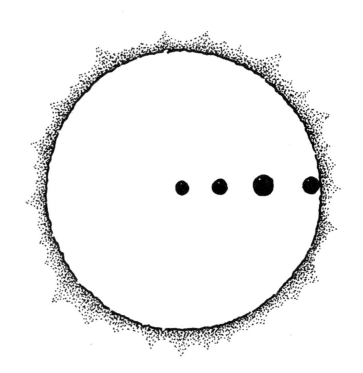

TRACKING THE DOG AND THE BULL

Orion is a good constellation to track from. Follow Orion's belt down to the brightest star in the sky, **Sirius**. This star that twinkles like a white diamond is often called the Dog Star because it is part of the constellation *Canis Major*, one of Orion's hunting dogs.

Sirius derives from a Greek word meaning "scorching." It burns at about 20,000° F (11,000° C) and has a −1.6 magnitude.

Wise ancient Egyptians carefully watched for the seasonal appearance of certain stars to predict events on Earth. Many of the Egyptian temples and pyramids were built to face the direction a particular star first appeared on the eastern horizon. They worshipped Sirius' first rising at dawn, believing that its appearance caused the annual flooding of the Nile River.

Gazing above and to the right of Orion, the belt stars will guide you to another red giant, **Aldebaran**. Aldebaran is the fiery right eye of *Taurus*, the Bull. Compared to Betelgeuse, Aldebaran has a rosy tone (you may need binoculars to see this color difference).

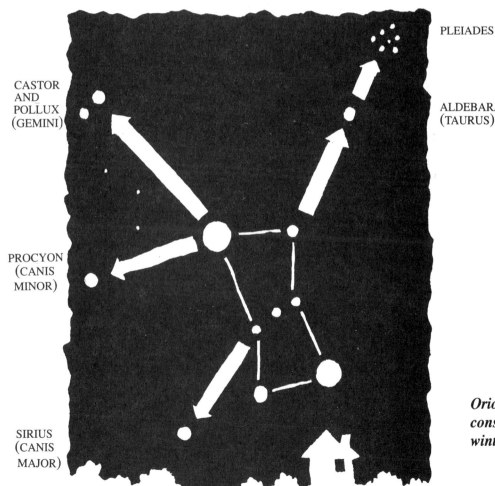

PLEIADES

CASTOR
AND
POLLUX
(GEMINI)

ALDEBARAN
(TAURUS)

PROCYON
(CANIS
MINOR)

SIRIUS
(CANIS
MAJOR)

Orion, as tracking constellation to the winter sky

Taurus, the Bull. Look below the bull's fiery red "eye" star, Aldebaran, for the Hyades cluster, and on its shoulder for the Pleiades cluster.

Do you see the "V" of stars made by following the horns down through the nose of the bull? The giant horns of Taurus were thought to open up the year and bring in springtime because it once appeared at the time of the spring equinox.

THE PLEIADES

Can you spot the shoulder of the bull? Look for a cluster of stars called the *Pleiades*. These stars are a good test for your eyesight. If it's really good, you'll be able to see a grouping of seven stars with your unaided eyes, although only six are easily visible. Binoculars will reveal dozens of stars, and a strong telescope will show hundreds of stars. Stars that were born together stay together for a while in space in an association known as a star cluster.

There are two types of star clusters, open and globular. Open clusters are relatively young groupings of stars. The Pleiades is an open star cluster that was born about 100 million years ago, during the Age of the Dinosaurs. The *Hyades* is another open cluster in Taurus that contains twenty-seven stars visible to the naked eye.

Although the Pleiades is a small constellation, it is the source of many legends and ceremonies. The Australian aborigines believed that the changes in the seasons must be due to the appearance of certain star patterns. Because they saw the sun every day, they didn't think the sun had anything to do with the summer's heat, but that the Pleiades did! They held ceremonial dances to honor these star gods.

At sundown on the night that the Pleiades was due to be directly overhead, the Aztec Indians of Mexico put out all fires. It was the night when the souls of their dead travelled to their last resting place. Exactly at midnight, a huge bonfire was lit at the top of an immense pyramid and someone was sacrificed in the flames. If the gods accepted the human sacrifice, the Pleiades crossed overhead without mishap, and the world was safe for another fifty-two years. If they did not, the Earth would be destroyed.

Aztec sacrificial ceremony. Long ago, the Aztec Indians of Mexico "saved" the Earth by offering a human sacrifice when the Pleiades passed overhead.

What spirits do you see in this group of stars? The Arabs pictured a group of camel riders, the Greeks spotted the seven daughters of Atlas, and some Native American tribes saw seven youths dancing. The story goes that the youths kept dancing and refused to come home for dinner. Later, when the angry parents refused to feed them, the youths danced up to the sky, where you can see them today.

Some Native American Indians see seven youths dancing in the sky.

The Spring Sky

The movement of the constellations is gradual—with patterns overlapping from one season into the next. As Earth moves from winter into spring, *Gemini*, the Twins, moves overhead in the night sky. Gemini takes the shape of a long, thin rectangle that is highlighted by two bright stars, **Castor** and **Pollux**. If you look at Castor through a telescope, you'll see that this one star is really a complex, six-star system. This star system contains three pairs of stars that revolve around each other, called *binary* stars. These three binaries revolve around each other as well.

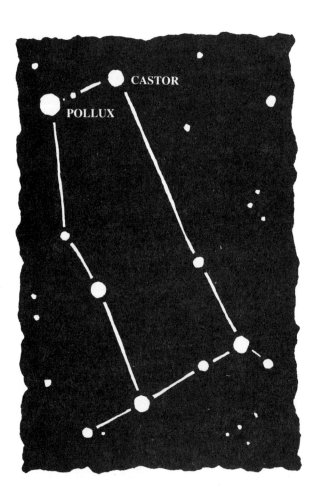

Gemini, with twin stars Castor and Pollux

By April (if Taurus has done his job properly), the spring constellations become prominent. The Big Dipper reaches its highest point in the northern sky and it helps us to detect three of the brightest stars in the springtime sky.

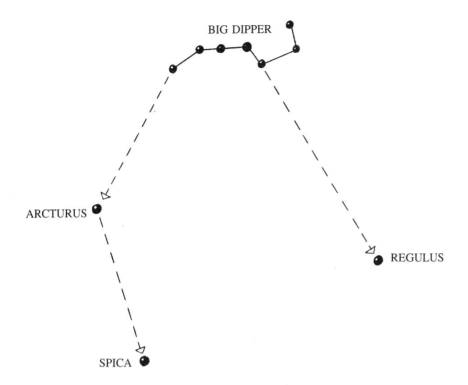

BIG DIPPER

ARCTURUS

REGULUS

SPICA

Tracking from the Big Dipper to the spring sky constellations. Follow the handle to Arcturus, in Boötes, and on to Spica, in Virgo. The bowl stars lead you to the blue-white star Regulus, in Leo.

LEO, THE LION

A line from the bowl of the Dipper (away from the North Pole) will take you to **Regulus**, which is part of *Leo*, the Lion. Regulus, which is Latin for "Little King," was named by Copernicus, the father of astronomy. Can you see the five stars that form a sickle or a backwards question mark? Regulus forms the "dot" of the question mark. Now, imagine these stars are the lion's head and chest. Look to the left of Regulus until you see a triangle of stars. These make up the lion's rear haunches and legs. The legend goes that Leo was famous as the King of the Beasts. No one could kill

Constellation Leo, the Lion

43

him. He had a hide so tough that arrows just bounced off his skin. Then one day Hercules, the Greek hero famed for his tremendous strength, crept up and stunned Leo by hitting him over the head with a huge log. Hercules then crushed the lion to death with his massive arms. To honor this mighty beast, the gods placed Leo among the stars.

OTHER CONSTELLATIONS

Coyote juggling his eyeballs

If you follow the curve of the handle of the Big Dipper, you'll reach a yellow-orange giant star called **Arcturus**. The Spokane, a Northwest Native American tribe, call this star "Coyote's Eyeball." The coyote is often portrayed as a trickster in Indian legends. One day, a coyote was showing off to a group of girls by juggling his eyeballs. He juggled one so high that it stuck in the heavens, where you can still see it today.

Arcturus is part of a kite-shaped constellation, *Boötes*, the Herdsman. Boötes' task is to guide Ursa Major and Ursa Minor as they circle around the North Pole.

Close to Boötes is a semicircle of faint stars that resembles its name, *Corona Borealis*, or the Northern Crown. The Blackfoot Indians believed that the crown was the lodge of the Spider Man, a god who spun his webs in the heavens. The webs would stretch down to the Earth and allow the heavenly spirits to travel back and forth between the Earth and the sky.

Northern Crown (left) and *Boötes (right)*. *Arcturus, in Boötes, is one of the brightest stars in the spring sky. The Northern Crown's semicircle of stars is easy to spot next to Boötes.*

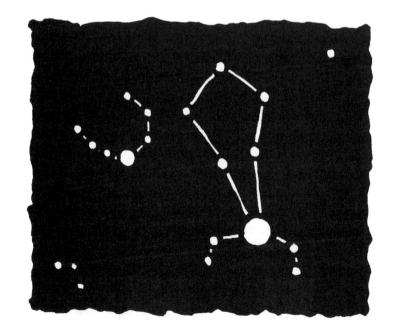

Now follow the curve through Arcturus down until you reach a brilliant blue star, **Spica**. (Memory trick: arc to *Arc*turus, then speed to *Sp*ica). Spica is part of the constellation *Virgo*, the Maiden. Virgo was the Roman goddess of Justice. Originally, she lived on Earth, but the wickedness of humanity drove her away to live in the heavens. Virgo is also considered the goddess of fertility. English farmers used to wait for Spica's first appearance in the spring sky before they planted their wheat in order to be sure of a good crop. Virgo stands high in the sky as spring turns into summer.

Farmers waited for Spica to appear before planting their crops.

The Summer Sky

Summer is often the time for travelling and relaxing under the stars in a mountain meadow or along a shoreline. Although the heavens are not packed with brilliant stars, there are three distinct stars to watch for. These widely spaced stars map out a summer triangle (or winter triangle for those living in Australia and New Zealand).

THE SUMMER TRIANGLE

Deneb is a hot, blue-white star in the constellation *Cygnus*, the Swan. Deneb means "tail" in Arabic and marks the tail of a graceful swan seen with its wings spread, as if in flight. Cygnus is also

The brightest summer star, Vega, in the constellation Lyra

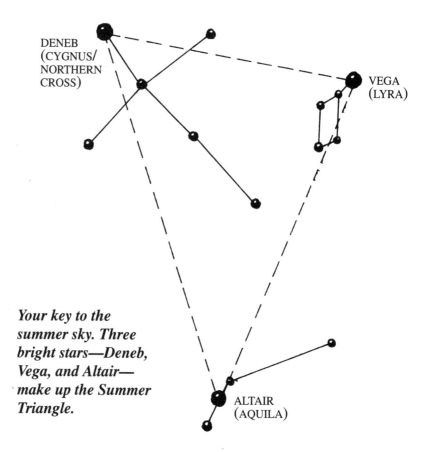

DENEB
(CYGNUS/
NORTHERN
CROSS)

VEGA
(LYRA)

Your key to the summer sky. Three bright stars—Deneb, Vega, and Altair— make up the Summer Triangle.

ALTAIR
(AQUILA)

called the Northern Cross. The Aleutian Indians of Alaska see neither a cross nor a swan but, instead, the spirit of a seal hunter in his kayak. What do you see?

South of Deneb is the brightest and bluest summer star, **Vega**. Vega is part of a small constellation, *Lyra*, the Lyre. The lyre is a musical instrument similar to the harp. Roman legend claims that the god Mercury created it by stretching strings across a tortoise shell. The Chinese call this constellation *Chih Neu*, the Celestial Weaver. This goddess oversees the weaving and painting of fine tapestries by Chinese women.

Completing the triangle is **Altair**, which means "the flier" in Arabic. Altair is a part of the constellation *Aquila*, the Eagle.

Aquila, the Eagle, with Altair on its shoulder

The constellation Cygnus, as a seal hunter in a kayak

DANGER STALKS THE SCORPION

By the middle of summer, sky detectives who live south of the equator get the best views of the constellation *Scorpius*, the Scorpion. In one legend, Scorpius stings and kills Orion, the Hunter. That's why you'll never see Orion when Scorpius is high in the sky. At the center of the fishhook shape of Scorpius is a supergiant, fiery red star, **Antares**. It is the "heart" of the Scorpion. Antares means "rival of Mars," and is often mistaken for the planet Mars.

Nearby, *Sagittarius*, the Archer, stands poised with an arrow ready to shoot the Scorpion. Sagittarius rises over the horizon towards the end of summer. Some people see a teapot instead of an archer in this star pattern.

Scan the skies between Scorpius and Sagittarius with binoculars for a kind of star group called a *globular cluster*. It looks like a white smudge in the night sky. Globular clusters are generally much older (more than 15 *billion* years old) than open clusters and are packed with millions of stars.

The constellations Sagittarius (left) *and Scorpius* (right) *with the red star Antares as its heart*

The Autumn Sky

The autumn skies are dim but wonderfully alive with legends of heroes and heroines. Cassiopeia sits boldly overhead to guide northern sky detectives through an action-packed, ancient Greek legend. Look below the "W" for a large square that's part of *Pegasus*, the Flying Horse. You'll probably need to stretch your imagination to see a horse flying upside down in these stars.

Cassiopeia, as tracking constellation to the autumn sky—Capella in Auriga to the left, and below to the square of Pegasus and on to Fomalhaut in the Southern Fish.

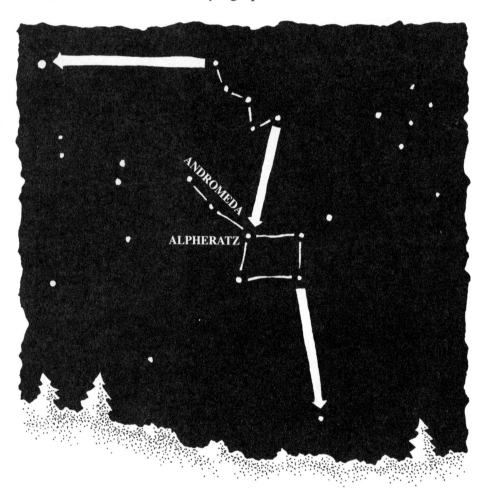

THE CHAINED PRINCESS

Attached to Pegasus (by the star Alpheratz) is Cassiopeia's daughter, *Andromeda*, the Chained Princess. As part of her punishment for bragging, Cassiopeia was being forced to sacrifice Andromeda to a sea monster. Fortunately, Andromeda was spotted by *Perseus* on his return from killing the dreaded creature *Medusa*. (Medusa's head was covered with snakes and her face was so horrid that anyone who gazed upon her turned to stone. Pegasus, the horse, sprang out of the blood of Medusa.) Perseus slew the sea monster, freed the princess from her chains, and married her.

THE "WINKING" STAR

Perseus is pictured in the sky holding the head of Medusa in his left hand. Can you locate **Algol**, the "demon" star that is Medusa's left eye? If you watch this star for a week, you'll see that about every three days it changes in brightness as if it is winking. The Arabs thought that this star was a sinister spirit with a wicked sense of humor. The word "ghoul," an evil spirit that robs graves and feeds on corpses, comes from the star's name, Algol.

From Greek legend: Perseus with the head of Medusa, and the Princess Andromeda

Algol is a binary star. One of the stars of this star pair is brighter than the other. When the smaller, brighter star (illus. **A**) passes in back of the larger, fainter one (illus. **B**) the total light coming to Earth from this binary drops temporarily, creating the "wink." Modern astronomers call this star set an "eclipsing binary."

Eclipsing binary. The star "winks" when the brighter star moves behind its larger, duller companion.

SIGNS OF COMING WINTER

Two sides of the square that forms Pegasus point to a constellation more visible to southern watchers, *Piscis Austrinus*, the Southern Fish. The slightly reddish star **Fomalhaut** forms the mouth of this fish. Long ago, its rising marked the winter solstice.

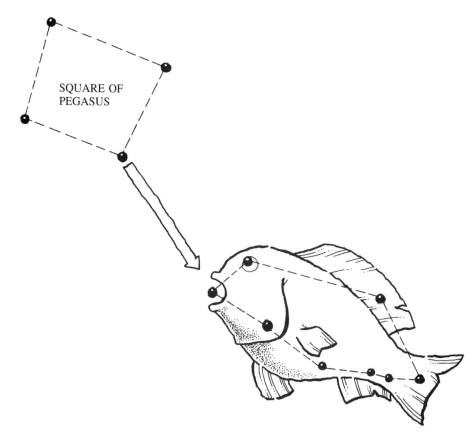

*Two sides of Pegasus "point" to Fomalhaut at the mouth of the Southern Fish, **Piscis Austrinus.***

Do you see a bright star that's yellow, like our sun? If you imagine a line from the start of Cassiopeia's "W" and go past Perseus, you'll reach **Capella**. Capella is part of the house-shaped constellation *Auriga*, the Charioteer. It heralds the coming of winter. This twinkling star can be seen as far south as New Zealand. The Chuckchee, a tribe in northeastern Siberia, see two reindeer pulling a sled in this constellation.

Auriga, as reindeer and sled

These are just a few of the many constellations that shine in the night sky. A detailed star chart for the latitude at which you live will help you detect some of the less conspicuous patterns and "spirits" that journey across the heavens.

Case #3: Ghost Hunt

What ghosts can you find among the stars? Using your star frame and lots of imagination, hunt among the stars for undiscovered ghosts and spirits that might be hovering in the night sky. On a sheet of paper, mark down the positions of the stars you capture with your frame and trace the outlines of the phantom you've discovered. Then, create a legend to go with it. You might tell about how a particular ghost came to live among the stars, why it haunts the other heavenly spirits, and what magical powers it has over the people on the Earth.

This is especially fun to do with a group of friends. Try this on different nights throughout the year and soon you'll have your own personal evening ghostbook.

Case #4: Binocular Hunt

Test your detective skills and hunt for star treasures. Use the chart to keep track of your nightly discoveries. Binoculars will help you uncover most of these hidden gems. If you don't have a set in your family, check with Army–Navy surplus stores. They often carry inexpensive field glasses. Look at the whole-sky maps here. They show all the constellations visible from the Northern and Southern Hemispheres. Turn the book until the present month on the map of your hemisphere is towards you. The stars you can see at that time of year will be in the lower half of the map. How many constellations you actually see will depend on how clear the sky is, the time of night, and your *latitude* or distance north or south of the equator. A *planisphere*, or star dial, for your latitude will show exactly what stars are visible in your sky for any hour of the year.

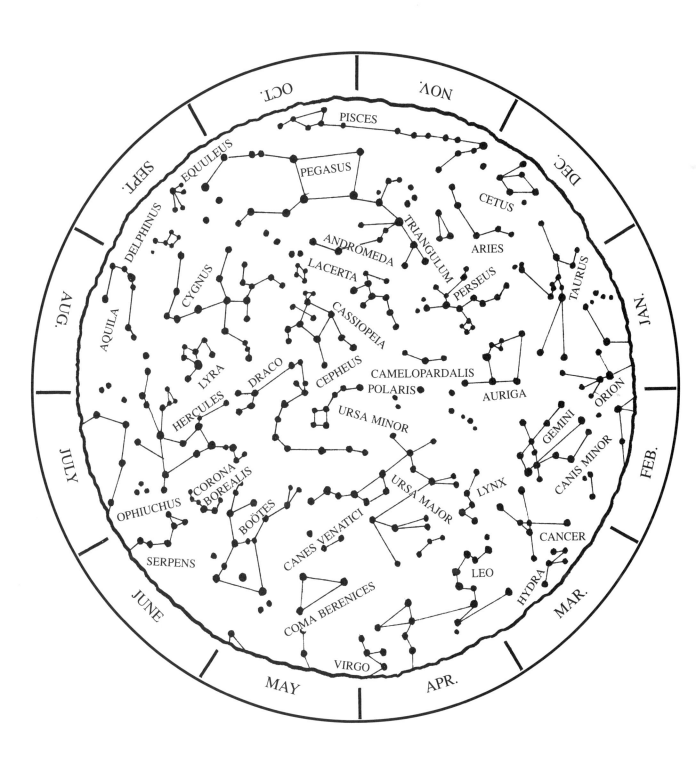

Whole-sky star map—Northern Hemisphere

53

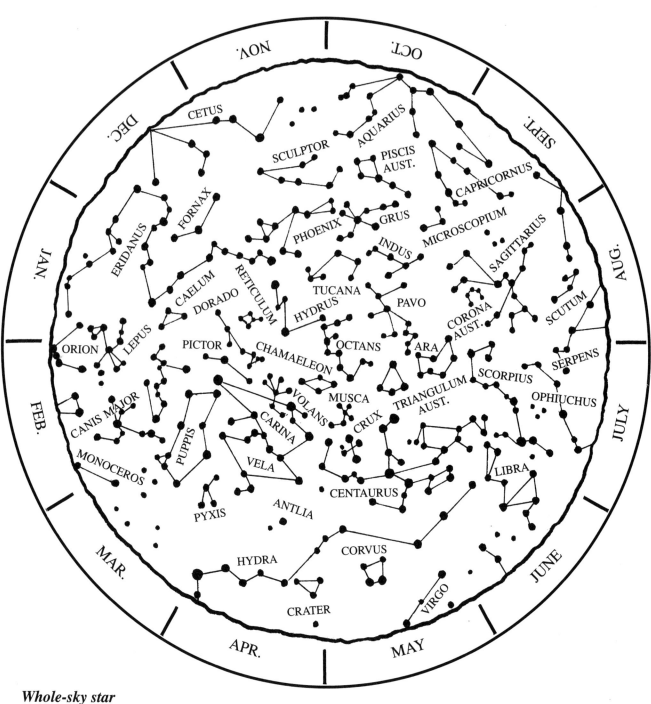

Whole-sky star map—Southern Hemisphere

54

I On a clear, dark night, search for stars that show a particular color. You might use a form like the one here to record your observations.

STAR COLORS

COLOR	STAR	CONSTELLATION
BLUE	Spica	Virgo
	Vega	Lyra
	Achernar*	Eridanus
BLUE-WHITE	Rigel	Orion
	Deneb	Cygnus
	Regulus	Leo
WHITE	Sirius	Canis Major
	Altair	Aquila
	Castor	Gemini
YELLOWISH WHITE	Canopus	Carina
	Procyon	Canis Minor
YELLOW	Capella	Auriga
YELLOW-ORANGE	Arcturus	Boötes
ORANGE	Atria*	Triangulum Australe
	Pollux	Gemini
RED	Aldebaran	Taurus
	Fomalhaut*	Piscis Austrinus
	Betelgeuse	Orion
	Antares	Scorpius
	Mira	Cetus
	Garnet Star	Cepheus

II Throughout the year, hunt in and around these constellations for open and globular star clusters.

STAR CLUSTERS

CONSTELLATION	CLUSTER TYPE
Cancer	open; called *Praesepe* or the Beehive; splendid views with binoculars
Gemini	open; brightest of all open clusters
Hercules	globular; called M13; finest in the northern sky—over 100,000 stars

Perseus and Cassiopeia	open; double cluster between these two constellations
Canis Major	open; called M41; look south of Sirius
Taurus	open; look on the Bull's face for the Hyades and on its shoulder for the Pleiades
Auriga	open; sweep through the center for a couple of beautiful clusters
Sagittarius and Scorpius	open and globular; scan the area between these two constellations
Centaurus*	globular; large, contains about one million stars
Crux*	open; called the Jewel Box; has one giant red star shining amongst many bluish stars
Tucana*	globular; a very large and bright cluster

* Southern Hemisphere stars and constellations

Case #5: Sky Jungle Stalk

I Many people imagined animals stalking through the night sky. Match up the clues with the animal constellations listed below:

CLUES

① hibernates on land but not in the sky

② this toothless creature has a powerful bite

③ was once an ugly duckling

④ has a bill that doesn't fit in your wallet

⑤ a hunter's helper and maybe your pet

⑥ gets a charge out of the matador's red cape

⑦ his mane is his pride, and his pride is his family

ANIMAL CONSTELLATIONS

ⓐ Cygnus, the Swan

ⓑ Canis Major, the Dog

ⓒ Ursa Major, the Bear

ⓓ Leo, the Lion

ⓔ Tucana, the Toucan

ⓕ Taurus, the Bull

ⓖ Scorpius, the Scorpion

Answers: 1. c, 2. g, 3. a, 4. e, 5. b, 6. f, 7. d

II   There are many other animals in the sky jungle. See if you can find the following animal constellations. You may need a sharp eye to spot these smaller, dimmer star patterns.

CONSTELLATION	TRACKING CLUES
Delphinus, the Dolphin	A tiny but bright constellation that looks like a diving dolphin. Look near the star Altair, in *Aquila*, in summer.
Corvus, the Crow	Curve down from the star Spica, in *Virgo*, to see this four-sided animal shape in spring.

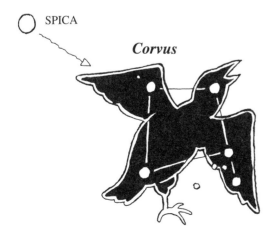

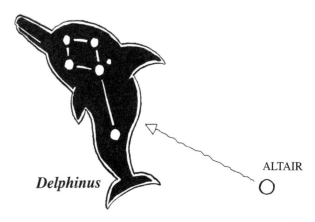

Aries, the Ram	Only two bright stars stand out in this fall zodiacal constellation. Search east of *Pegasus*.
Grus, the Crane	This group of stars looks like a bird in flight with outstretched wings. Southern detectives will spot the crane near the star Fomalhaut, in *Piscis Austrinus*, in spring.
Lepus, the Hare	This animal crouches below the star Rigel, in *Orion*, in winter. Look for two stars that make up the hare's ears.

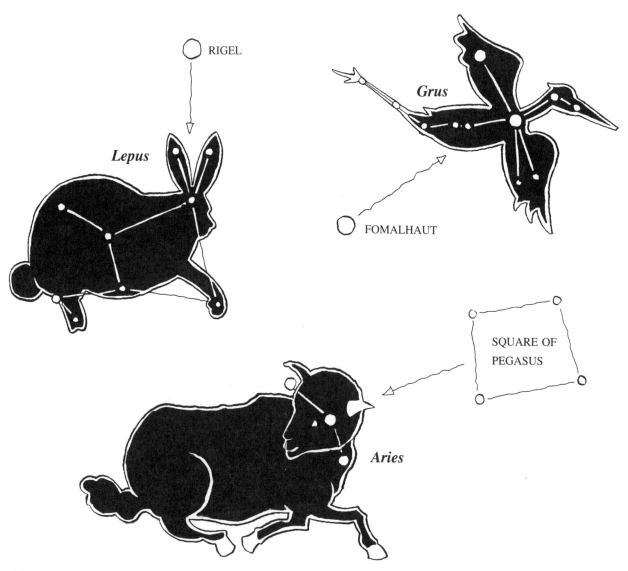

•4•
VANISHING ACTS

The moon, with its ever-changing shape, has fascinated and frightened people since the earliest times. Temples were erected, prayers were chanted, and dances were performed to honor this mysterious body. When space travel fulfilled the seemingly impossible dream of a moon landing, many riddles about Earth's closest neighbor were answered. Today, the moon still retains its romance, and its rugged landscape captures our curiosity. Watch closely—not only does the moon change its appearance every night, it occasionally pulls a surprise vanishing act.

Spying on the Moon

When you step outside on a clear night, look to see if there's a moon overhead. The moon is an easy object to spy on because it's bright and large. If you look for the moon every night, or every few nights, for a week or so, you'll notice that its shape is always changing. Some nights it looks like a fingernail clipping, other nights it's a glowing globe, and sometimes, it disappears com-

Night after night, the moon changes shape.

pletely from view. If you check on the moon each night at the same time, you'll also find the moon in a very different position in the sky from one night to the next. (It will be in different positions in relation to the constellations, too.) What causes these changes?

THE MAN IN THE MASK

The Innuits, natives of the Arctic, believe that the different moon shapes are masks worn by the Man in the Moon. Long, long ago, the Man in the Moon had just one mask, a large round one with a sad face painted on it. It was a very heavy mask and he had to use all his strength to hold it up with his hands. He was very lonely, and tired and unhappy about his burden. One night, as he looked down on the Earth, he noticed a young girl, Nesti, admiring the beauty of the moonlight by the seashore. Nesti called to the Man in the Moon. She said that she loved him and wanted to live with him in the heavens. The Man in the Moon told her to think about it for five days because the work would be hard and they'd have little time together. When he returned, she surprised him with masks

The Man in the Moon

of different sizes that she had made out of silver fox furs. This made the Man in the Moon very happy, and he gathered Nesti and the masks up into the heavens. Today, you can see the Man in the Moon and Nesti sharing the job of carrying the masks across the sky. He carries the bigger half and full moons, and she carries the smaller crescent masks.

MOONGLOW

Unlike the stars, the moon has no light of its own. It shines by the sunlight that bounces, or reflects, off of its surface. The moon is the Earth's natural satellite, the only heavenly body that revolves about the planet. It circles the Earth once every 29-1/2 days. During its trip, the moon seems to change shape, or go through

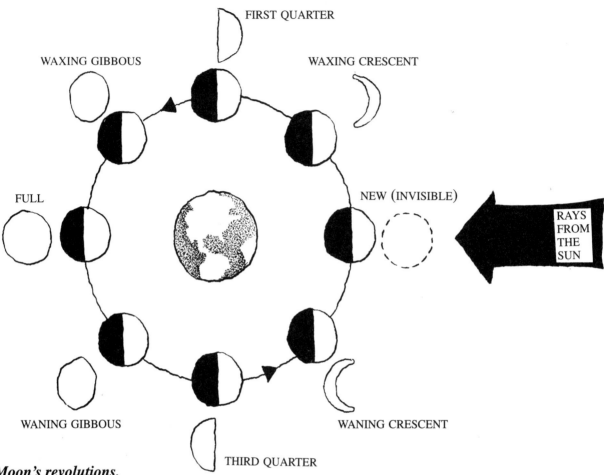

FIRST QUARTER

WAXING GIBBOUS

WAXING CRESCENT

FULL

NEW (INVISIBLE)

RAYS FROM THE SUN

WANING GIBBOUS

WANING CRESCENT

THIRD QUARTER

Moon's revolutions. As it circles the Earth, the moon seems to change shape, called going through phases. The outer shapes show the phases of the moon as seen from Earth.

phases. These phases are caused by the changing positions of the moon relative to the Earth and the sun. The moon's revolution about the Earth is also why you see the moon in different places in the night sky.

DISAPPEARING ACT

Some nights the moon is high in the sky just after sunset. On other nights it doesn't appear until after midnight. Sometimes when you go hunting for the moon on a clear evening, you can't find it. That's because one night during each revolution, the moon disappears. It is still there, of course, but it is invisible. You can't see it because the part of the moon lighted by the sun is facing away from you. This phase, called the *new moon*, is when the moon lies between Earth and the sun. The stars really sparkle on a clear, new-moon night.

Seeing the Moon's Phases

To solve the mystery of why the moon is always changing its shape:

① Hold an orange at arm's length towards a bright lamp (illus. **A**).

② Make sure there's no other light in the room.

③ Pretend the lamp is the sun, your head is the Earth, and the orange is the moon moving around the Earth.

④ Standing in place, turn slowly around to the left, so that the orange moves around you in a complete circle.

⑤ Watch how the lighted part of the orange (moon) changes and seems to go through phases as you turn (illus. **B**).

MOTHER MOON

Many tribes around the world believed the moon to be something mystical. In dozens of legends, the moon is a goddess, often the wife or sister of the sun. Together with the sun and the stars, the moon directed the timing of people's daily lives. The Pawnee Indians saw the moon as a powerful female spirit whose job it was to take care of her daughter Earth. The first new moon in spring, after the leaves appeared on the willow branches, was important to the medicine men of the tribes. This was the time that the moon imparted special powers to these healers.

Pawnee medicine man dances for healing power under the new moon.

The Phases of the Moon

Sometimes the moon has a thin crescent shape. If you observe it carefully at this phase, after sunset, you'll be able to see the dark part of the moon illuminated by a faint light. That light is called *Earthshine*. Earthshine is light that has travelled from the sun to the Earth, then from the Earth to the moon, and back again to the

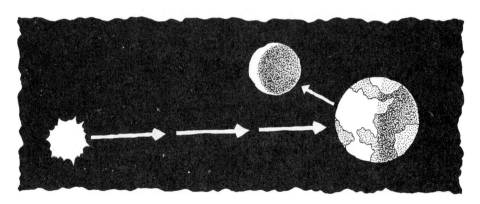

Earthshine on moon. Sunlight, reflected from the Earth, lights up the darkened part of the crescent moon.

64

Earth. Some people refer to this phase of the moon as "the old moon in the new moon's arms."

About seven days after the new moon, the moon grows, or *waxes*, to become a half-circle. This is the *first-quarter* phase, and it shows that the moon has completed one-fourth of its journey around the Earth. Look for this half-moon in the early night sky.

First-quarter phase

The moon continues to wax into a *gibbous* moon. Gibbous means humpbacked. The gibbous moon seems to be swollen on one side.

Fourteen or fifteen days into its cycle, the moon becomes *full*. Now, the moon sits farthest from the sun and you see its whole face glowing brightly in the black sky. This is the most dramatic phase. The night of the full moon is often the setting for ghost stories and horror scenes. Many legends say that, during a full moon, werewolves howl and vampires and lunatics roam the Earth. Some people believe that, if the full moon shines on someone who is sleeping, that person will go mad. The word lunatic comes from the Latin work for moon, *luna*.

Waxing gibbous moon

SECRET OF THE SHRINKING MOON

Have you ever watched the full moon rise above the horizon just after sunset? It looks so large. Then, as the full moon moves higher over the Earth, it appears to get smaller. Is the moon shrinking? The ancient Greek astronomer Ptolemy figured out the solution to this puzzle. The difference in the size you see is really only a trick your mind plays on you. When the moon is near the horizon, it

looks large because your brain is comparing the moon to the familiar background of trees and houses. Because of the rotation of the Earth, the moon then appears to rise higher in the sky. When if is overhead, there are no clues to judge size by, so your mind tells you that the moon looks smaller. Also, for some reason, an object you see overhead appears smaller than it does when you see it straight ahead on the horizon. You'll notice the same optical illusion occurs with the rising and setting sun.

Eclipses

If you're a careful sky detective, you may be able to catch the full moon pulling a surprise vanishing act. A darkness will slowly slide over an edge of the moon and creep across its face until the moon finally disappears. You can imagine how such an event would frighten the people of long ago. It did not happen very often, so when it did people thought that it was happening because the gods were angry at them for some reason. They were afraid that the moon was leaving the sky forever. The Cherokee Indians of North America claimed that a huge frog was eating the moon. They would beat kettles, ring bells, and yell until they frightened the frog away and saved the moon.

Frog eating moon

66

LUNAR ECLIPSE

What you're really watching when you see the moon disappear like that is an *eclipse*. When the sun shines on the Earth, the Earth casts a long cone-shaped shadow out into space. A *lunar eclipse* occurs when the moon moves into the deep shadow of the Earth where it seems to disappear. If you look closely, you'll see that the moon never totally vanishes from the night sky. Instead, it glows a deep copper color. This is because the Earth's atmosphere scatters some of the sun's light onto the moon.

Lunar eclipse—when the moon enters Earth's shadow during certain **full-moon** *phases*

Do you know why you don't see an eclipse every full moon? That's because the moon's path is tilted, so it usually passes above or below the Earth's shadow. Only when the moon moves into the *umbra*, the darkest part of Earth's shadow, do you see a total eclipse. Often, the Earth's shadow just clips the edge of the moon and you get a less dramatic partial eclipse.

SOLAR ECLIPSE

The moon also produces a long shadow in the sun's light. When the Earth passes through the moon's shadow, we experience a *solar eclipse*. Solar eclipses can occur only when the moon is new,

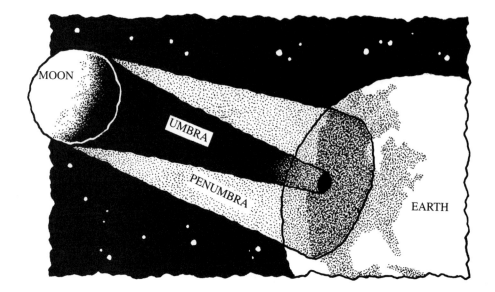

Solar eclipse—when the Earth enters the moon's shadow during certain new-moon phases

MOON

UMBRA

PENUMBRA

EARTH

and there are at least two solar eclipses a year. Each solar eclipse is seen only in a small area of our planet because the moon's shadow is less than 100 miles (160 km) wide when it strikes the Earth. A lunar eclipse, though, can be seen everywhere on Earth.

Many Moons

After the full-moon phase, watch how the moon's face shrinks in size, or *wanes*. The phases go from *gibbous* to *three-quarter* to *crescent*, until the moon fades to a new moon. In western China, the Moslems believed that Allah cuts up the waning moon into little pieces and makes stars out of it.

Waning gibbous moon   *Third-quarter moon*   *Waning crescent moon*

The moon rises approximately fifty minutes later each night. To catch the waning three-quarter moon, you'll have to get up in the middle of the night.

The changing shape of the moon was the basic guide for the earliest calendars. People paid attention to how the moon would wax and wane, and when it would rise and set. They understood that it took about 29 days to go from one new moon to the next and that a year would have 12 and, sometimes, 13 "moons." Several of the "moons" were given names according to the cycles of nature. Some Native Americans called the time when mosquitoes were thick the "Moon of Blood." The "Wolf Moon" would shine in late winter, when the wolves were hungry and cold and travelled in packs, bringing danger to people and other animals.

Wolf moon

Today, many events are still timed by the moon's cycle. The dates of Passover, Easter, Ramadan, and Chinese New Year are calculated according to certain full moons. The full moon nearest to the September equinox is frequently referred to as the Harvest Moon. The extra light given by this especially bright moon helped farmers work late into the night to gather in the last of their major crops before winter arrived. In China, this autumn moon ushers in a joyful celebration honoring the family. Its members travel home to join their relatives and eat a festive meal, which includes specially prepared mooncakes.

Moon festival. Mooncakes are part of the fun, when the Chinese celebrate the family during the time of the autumn full moon.

Moonbow

Can you spy a moonbow? If you look for the moon when there is a high layer of clouds in the night sky, you'll see a colorful arc or circle around it. The moonbow is usually red on the inside, and yellow and blue on the outside. It forms when moonlight is *refracted*, or bent, by ice crystals inside the high, thin cirrostratus clouds. When you see clouds cover the moon, they are Earth's clouds. The moon has no atmosphere, so has no moisture for making clouds. With no protective atmosphere, temperatures on the moon can reach 270° F (134° C) during the moon's day and cool to −270° F (−170° C) on the nighttime side!

Moonbow. The moisture in Earth's atmosphere sometimes creates a halo or moonbow around the moon.

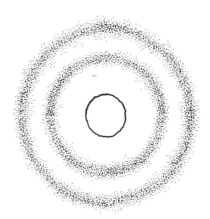

GRAVITY

You may wonder, what keeps the moon circling the Earth? Gravity, a force that acts much like a huge magnet, keeps the moon in its *orbit*, or circle, around the Earth and keeps them both in a path around the sun. The Earth's gravity also prevents you from flying off into outer space. The moon has its own gravity but

Because of its weaker gravity, astronauts bounced on the moon.

its pull is only one-sixth of that felt on Earth. This is because the moon is smaller than the Earth (about the size of a golf ball compared to a softball). The force of gravity becomes stronger as an object becomes bigger and more dense. When the astronauts walked on the moon, they were much lighter than they were on Earth and they bounced off the ground. To find out what you'd weigh on the moon, take your weight and divide by six.

PICTURES IN THE MOON

Almost every person who gazes at the face of the moon claims to see some sort of animal or person there. The Shan people of Burma see a silver hare. Malaysians see an aged hunchback sitting under a tropical banyan tree braiding strands of tree bark to make a fishing line. In Siberia, the Yakuts glimpse a girl carrying on her shoulders a pole with two buckets. What do you see?

The Shan people see a hare in the moon.

The Lunar Surface

The dark patches of the moon that create the imaginary shapes were first observed by the Italian astronomer Galileo in 1609. He believed these large flat areas were oceans and called them *maria*, the Latin word for seas (singular, *mare*). But we now know that

Galileo and his spotting tubes. The new device (first telescope) helped Galileo explore the heavens and make many astonishing discoveries.

there's no water on the moon. Instead, the maria are dust-covered deserts formed from ancient lava flows. Some of the maria stretch 100 miles (160 km) across and are easily visible to the naked eye, especially during the first- and last-quarter phases. It was on the Mare Tranquillitatis that astronaut Neil Armstrong took the historic first moon walk on July 20, 1969.

Astronaut Neil Armstrong on the moon's Sea of Tranquillity

Without water to drink or oxygen to breathe, the moon isn't a place anyone would want to call home. But with its beautifully textured surface, it is a wonderful sight for sky detectives. A pair of binoculars will help you uncover the moon's spectacular geography. The moon's lighter regions are called the *highlands*. They are riddled with *craters*, huge, saucer-like depressions. Some of the craters were formed by the eruption of ancient volcanoes. Most of them, however, were made by meteors and other material from outer space bombarding the moon's surface. The craters you see are the scars of these ancient impacts.

Meteor forming crater. The entire lunar surface is marked by craters, most formed by such crash landings.

MANY KINDS OF CRATERS

The thousands of craters on the lunar surface come in a variety of shapes and are named after early astronomers and philosophers. The largest craters are only found on the highlands. You might spot some with mountain peaks in the center, terraced with a series of flat surfaces, or with chains of smaller craters snaking across their floors. Still others are split by long narrow furrows called *rills*. While most are cup-shaped, a few are hexagonal (six-sided),

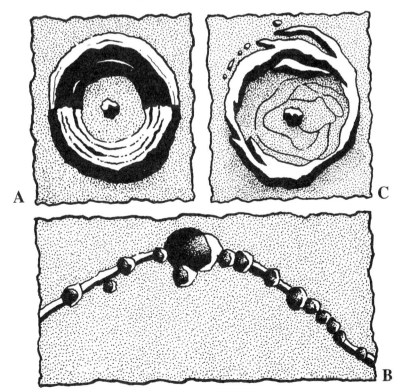

Crater types: terraced crater with mountain in center (A), crater chain (B), and rilled crater (C)

like Ptolemaeus, which lies near the center of the moon. Clavius is the largest crater. It has a diameter of 135 miles (225 km). In the Mare Procellarum, the Ocean of Storms, is a crater called Copernicus. Although it's only 50 miles (80 km) wide, its walls are 17,000 feet (5.1 km) high. In its center is a mountain group with peaks rising to a majestic 22,000 feet (6.8 km), a little taller than North America's highest mountain, Mount McKinley.

When you investigate the lunar landscape, remember that you're always looking at the same side of the moon. As the moon circles the Earth, it also rotates or turns on its axis. The periods of rotation and revolution are so close that only one side of the moon ever faces the Earth. Before the voyages to the moon, no one had ever seen the far side. We now know that the far side of the moon is almost completely covered with craters.

Moon circling Earth. Because the moon turns once on its axis for every orbit of Earth, the same side is always facing Earth. See the position of the "moon spot" in this drawing.

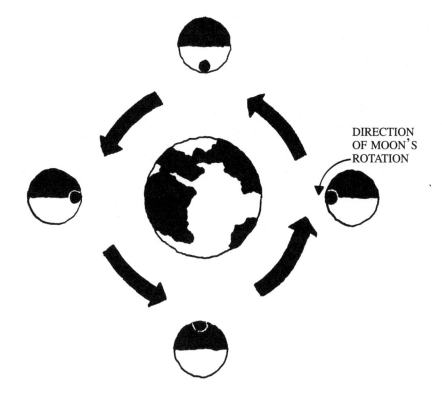

DIRECTION OF MOON'S ROTATION

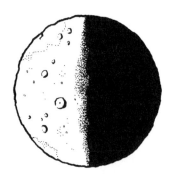

Terminator, dividing line between light and darkness

Can you find the *terminator*? This is the boundary between the dark and lighted portions of the moon. During the crescent-moon phases, craters and mountain chains stand out boldly along the terminator.

Try focusing your binoculars on the moon's Southern Hemisphere during a full moon. You'll see the crater Tycho with

Tycho crater, showing rays

an impressive system of lines, or *rays*, that stream out for 625 miles (1000 km). Tycho is a relatively young crater, only a few hundred million years old. At that time, this area of the moon's surface was hit by a huge meteor. Large amounts of dust and rock splashed out from the crater and fell back to the surface as the rays you see today.

Craters look very much as they did when they were formed millions and millions of years ago. Because the moon has no atmosphere, there's no water or wind to erode its surface. How the moon was formed is still an unsolved mystery. *Selenologists*, scientists who study the moon, have many different theories about that. They agree, however, that the moon and the Earth are about the same age (4.6 billion years old), are made up of the same basic minerals, and were formed at the same time.

Case #6: Lunar Probe

Probe the moon's rugged landscape as it goes through its monthly cycle of changes. Use the map here to help you scout for craters, maria, and mountains with binoculars. What you see in the map is visible on the moon with just your naked eyes or with binoculars. If you are using binoculars, prop your elbows up on a fence, the arms of a chair, or set the binoculars on a tripod. This will help you to see better and not get tired. If you look at the moon with a telescope, remember that the features will be upside down.

To use the spotting chart below, start with the new moon phase, day 0. The actual days listed are only an approximate guideline. Generally, you can see the elevated walls of the craters best between the crescent and quarter phases. When the moon is full the light is "flat" and you don't have the shadows that make the craters stand out.

DAY	PHASE	FEATURES
1–4	Thin Crescent	Look now for Earthshine; Mare Crisium and crater Cleomedes on the moon's northern edge; Mare Foecunditatis coming into view.
5	Waxing Crescent	Mare Tranquillitatis—site of first moon walk; probe for three craters, Theophilus, Cyrillus, and Catharina on this mare. The walls of the terraced crater Theophilus rise to 21,000 feet (6.4 km).
7	First Quarter	Great string of craters along the terminator; on western shore of Mare Serenitatis look for Haemus and Caucasus mountain ranges.
9–11	Waxing Gibbous	A giant crater, Clavius, at South Pole; in Mare Imbrium—greatest lunar chain, Apennines Mountains, and Copernicus crater, the Monarch of the Moon. Look for Jura Mountains along the terminator, near the North Pole.
14	Full	Crater Tycho at South Pole with fantastic ray streaks; moon's brightest feature is visible—crater Aristarchus in Oceanus Procellarum. Look for the Man in the Moon, Moon Goddess, the Shan people's silver hare, etc.
15–18	Waning Gibbous	Huge maria surfaces: Mare Nubium and Imbrium, Sinus Iridum, Oceanus Procellarum; also Plato—a dark-floored crater near Mare Frigoris.

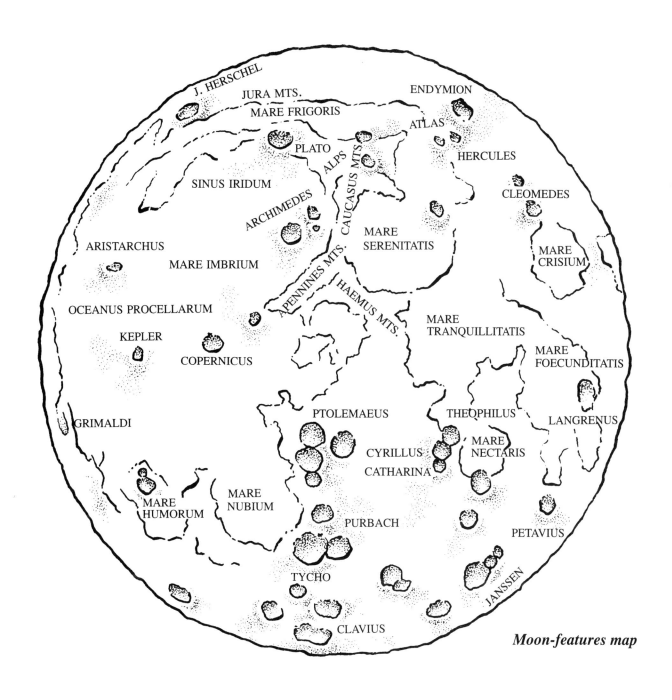

Moon-features map

| 21 | Third Quarter | Dark flat plains of Western Hemisphere still dominant; crater Ptolemaeus at terminator. |
| 23–28 | Waning Crescent | Oval, dark patch of Grimaldi crater on western edge; surface details disappear. Sinus Iridum at terminator. |

Name Meanings of the Major Maria

Mare Crisium: Sea of Crises
Mare Foecunditatis: Sea of Fertility
Mare Nectaris: Sea of Nectar
Mare Tranquillitatis: Sea of Tranquillity
Mare Serenitatis: Sea of Serenity
Mare Frigoris: Sea of Cold
Mare Nubium: Sea of Clouds
Mare Humorum: Sea of Moisture
Mare Imbrium: Sea of Showers
Sinus Iridum: Bay of Rainbows
Oceanus Procellarum: Ocean of Storms

Case #7: Vanishing Bodies

I Try to "catch" some of the sun's shadow as it vanishes during a solar eclipse. The safest way to do this is to make a pinhole projector. You'll need two pieces of white cardboard, each about a foot (30 cm) square, and a nail.

To catch and see a solar eclipse:

① Push the nail through the center of one piece of cardboard (illus. **A**).

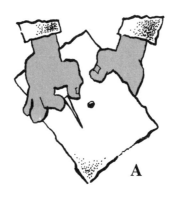

A

② Stand with your back to the sun.

③ Hold the board with the hole at your shoulder, so that the sunlight will shine through the hole.

④ Hold the second board about an arm's length away from the first board.

⑤ Move both pieces of board until you see an image of the sun projected on the second board (illus. **B**).

⑥ Watch how the sun's shadow "vanishes" during the eclipse.

This breathtaking event usually lasts only for a couple of minutes, but it can last as long as seven and a half minutes.

SCHEDULED TOTAL ECLIPSES OF THE SUN

DATE	WHERE VISIBLE
June 30, 1992	South Atlantic
Nov. 3, 1994	Peru, Brazil, South Atlantic, Pacific
Oct. 24, 1995	Iran, India, East Indies, Pacific
March 9, 1997	Mongolia, Siberia, Arctic
Feb. 26, 1998	Pacific, Columbia, North America, Venezuela
Aug. 11, 1999	North Atlantic, England, Central Europe, India

Warning: Just as you should never look directly at the sun, watching a solar eclipse through a dark glass or through pieces of film is **not safe**. During an eclipse, it only takes a few seconds to severely damage your eyes.

II Watch the moon vanish during a predicted lunar eclipse. It's safe to stare directly at the moon because its light is only reflected sunlight and cannot injure your eyes.

SCHEDULED LUNAR ECLIPSES

DATE	TYPE OF ECLIPSE
June 15, 1992	Partial
Dec. 9–10, 1992	Total
June 4, 1993	Total
Nov. 29, 1993	Total
May 25, 1994	Partial
April 15, 1995	Partial
April 3–4, 1996	Total
Sept. 27, 1996	Total
Mar. 24, 1997	Partial
Sept. 16, 1997	Total
Sept. 28, 1999	Partial
Jan. 21, 2000	Total
July 16, 2000	Total

Check an almanac, an astronomy magazine, or newspapers when the date is near, for the exact times of the eclipses.

•5•

VISITORS FROM OUTER SPACE

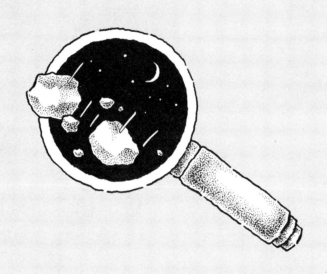

Fiery explosions, blazing streaks, lights that glow and crackle—are these invading aliens? You won't need helmets or protective gear to investigate these visitors from outer space—only a clear, moonless night. While some of these callers flash in the night sky like fireworks, be on the lookout for others that silently creep among the stars. And beware of black holes, the darkest and most mysterious of all space objects.

Shooting Stars

Zip! Suddenly what looks like a star shoots rapidly across a small area of the night sky. If you blink, it's gone. Could it be a light from an alien spacecraft? Probably not. Instead, these bright streaks of light are *meteors*. Meteors are commonly called "shoot-

ing stars," although they aren't really stars at all. Meteors are bits of space rock or dust that plunge into the Earth's atmosphere and burn up. The friction that's caused when these space particles speed through the atmosphere creates such heat that they glow and produce the trail of light you see. You'll probably spot about three or four "shooting stars" every night you sky-watch.

Every once in a great while, a large piece of cosmic dust will crash through Earth's atmosphere with a flash of light that lasts for five to ten seconds. These awesome *fireballs* are meteors that are so bright (brighter than the moon) they can even be seen during the daytime. Sometimes, the flash is followed by a loud bang. In the predawn sky of March 7, 1991, people in the northeastern United States were startled by a brilliant fireball that lit the sky for five seconds. Witnesses described it as "the size of a small car," "the shape of a boomerang on its side," and having "a red-and-green tail." Although startling and scary, meteors and fireballs are quite harmless invaders.

Showers of Meteors

Do you enjoy watching fireworks? If so, be sure to mark your calendar to see the next sky show of *meteor showers*. During a shower, dozens of meteors fall from the same place in the sky. Meteor showers occur when the Earth enters the dust trails left along the paths of certain comets. The showers are predictable because the Earth cuts through these comet trails the same time each year.

Earth passing through space dust. Stargazers see the dust, possibly left by a passing comet, as a meteor shower.

Meteor showers are named after the constellation you look at to see them. For example, the Leonids, a meteor shower that takes place around mid-November, has the constellation Leo in the background. Sometimes the dust particles are packed more closely along a portion of a comet's orbit and, when our planet crosses these swarms, you can see fantastic showers. This happens with the Leonids every 33 years: the next big shower should occur in November 1999.

People frightened by "falling stars"

The most spectacular meteor shower ever written about occurred on November 12, 1833, when 30,000 meteors per hour were reported to have "rained" down. People throughout North America thought all the stars were falling out of the sky.

Space Rocks

Should you put on a safety helmet or hard hat for meteor watching? When these dust pieces strike the Earth's atmosphere, they're travelling very fast, some 44 miles (70 km) a second. Most of the matter that makes up a meteor is vaporized when it enters the Earth's atmosphere.

A small amount of meteor material, however, releases a tremendous amount of energy. One ounce (28 grams) of meteor material has the energy of 30 pounds (14 kilograms) of TNT, but it is released mostly as light. Most meteors are smaller than a pea and burn up before they ever reach the ground. No person has ever been killed by the few that do strike the Earth's surface.

Once they land on Earth, these fragments from space are called *meteorites*. Meteorites have been found on every continent. They are named after the place where they are found after crashing to Earth. Most weigh only a few pounds, but some are huge. One that landed in southwestern Africa weighs 60 tons and remains wedged in the ground where it fell. Some tremendous craters have been produced from ancient meteorite impacts, such as Wolf Creek Crater in Western Australia and Winslow Crater in Arizona. The Winslow Crater was probably created about 25,000 years ago (during the last advance of the Ice Age glaciers), and it measures one mile (1.6 km) across and 600 feet (180 m) deep.

IS IT A METEORITE?

How would you know if you found a meteorite? You're very lucky if you happen to spot one hitting the Earth. Otherwise, look carefully at any rock you suspect might be one. A fresh meteorite will be covered with a black or reddish crust that formed during its fiery descent. Underneath, you'll find a grey or silver metal or a stony-looking material that will stick to a magnet. That's because all meteorites contain the metal iron, which is magnetic.

The early Eskimos searched for these hard space rocks and

Finding a meteorite

Winslow Crater. This huge crater in Winslow, Arizona, was formed when a meteor crashed to Earth long ago.

made knives out of them. The ancient Chinese believed meteorites were messengers from the gods. They feared that if left unburied, the stones had the power to ruin their crops. The Plains Indians also believed meteorites had supernatural powers. When going to war, the warriors carried bundles containing rocks they thought were meteorites to help them win their battles.

Other Sky Lights

SATELLITES

Another mysterious light you may see while scouting the heavens looks like a bright star moving rapidly across the entire sky. This is one of the thousands of artificial satellites orbiting Earth. It takes about five minutes for one of these to cross the sky.

Artificial satellites leave light trails.

Satellites appear to flash on and off as they tumble in orbit and sparkle from reflected sunlight. These spacecraft have helped to solve many of the mysteries about the night sky, but they have also created problems. Discarded rocket stages and burnt-out weather, communication, and military satellites are polluting outer space above our planet. Occasionally, this debris will shower down as people-made meteorites.

THE AURORAS

Do you live in the Far North or Far South, in places like Scotland, Alaska, New Zealand, or Chile? Then you'll be familiar with the *Aurora Borealis*, the Northern Lights, or the *Aurora Australis*, the Southern Lights. These eerie light shows compete with the special effects often seen in science-fiction movies. The auroras (named for the goddess of dawn) appear in many forms and intensities. A terrific range of brilliant greens, blues, oranges, and reds dance upwards from the horizon. Sometimes they resemble pulsating curtains of light that may even crackle and pop; at other times they

Aurora Borealis. Auroras are spectacular displays of light near the North and South Poles.

quietly appear as arcs that curve upwards in the sky like rainbows. Auroras are produced when electrically charged atomic particles from the sun collide with the gases in the Earth's upper atmosphere. The Earth's magnetic field channels these electrons and protons towards the North and South Poles, where the displays are most visible. Auroras are often seen around the times of solar flares, brilliant eruptions on the sun's surface. Major auroras have such a strong electrical charge that they disrupt radio communications and power lines.

According to the Naskapi Indians of Labrador, these dramatic lights are the spirits of dead people dancing. The Makah Indians of the British Columbia coast believed that the strange lights came from the cooking fires of a tribe of dwarfs who lived many moons'

Dwarf making the Northern Lights

journey to the north. The legendary dwarfs were evil spirits who stood no taller than the blade of a canoe paddle. Small but strong, they could catch whales with their bare hands. The lights of the auroras came from the fires lit by the dwarfs to boil whale blubber in big pots placed on the ice.

SKY CONE

Tropical sky detectives should keep a watch out for the *zodiacal light*. This is a faint cone-shaped light that appears along the horizon usually at or near the equator. Sometimes called the "false dawn," the zodiacal light is caused by sunlight reflecting off dust floating around in space. If you live in tropical latitudes, look for it along the western horizon after sunset in early spring, or along the eastern horizon before dawn in early autumn.

Zodiacal light. This faint cone-shaped glow is often seen after sunset and before dawn by watchers in the tropics.

Comets

If you scan the skies just after sunset or shortly before dawn, you may spy one of the most stunning visitors from outer space, a *comet*. Comets are city-sized chunks of gas, dust, and water vapor frozen into "dirty snowballs."

The appearance of a comet often seemed to signal a historical event on Earth. The Aztecs called them "stars that smoke" and believed that the fall of their ruler, Montezuma, was predicted by two comets in 1510. According to an Aztec legend, the god Quetzalcoatl had disappeared long ago into the west, promising to return someday out of the eastern sea to reclaim his empire. When

Montezuma seeing twin comets

Montezuma observed the two brilliant comets in the sky at the same time, he took them as a sign that Quetzalcoatl was on his way. Nine years later, when the Spanish conqueror Cortez arrived, Montezuma believed that the prophecy was fulfilled and surrendered his rule.

Napoleon's crushing defeat at Waterloo has even been blamed on the Comet of 1811 four years earlier.

HOME OF THE COMETS

Billions of comets exist in a belt of clouds on the outskirts of our Solar System, far beyond the planet Pluto. Sometimes a comet gets pulled out of the clouds by the sun's strong gravity and starts on a very long *elliptical*, or cigar-shaped, orbit around the sun. As it nears the sun, the comet's surface ice melts, releasing gas and dust that form a hazy cloud, or *coma*. When a comet ventures even closer to the sun, it stretches out and forms a ghostlike *tail*. (The

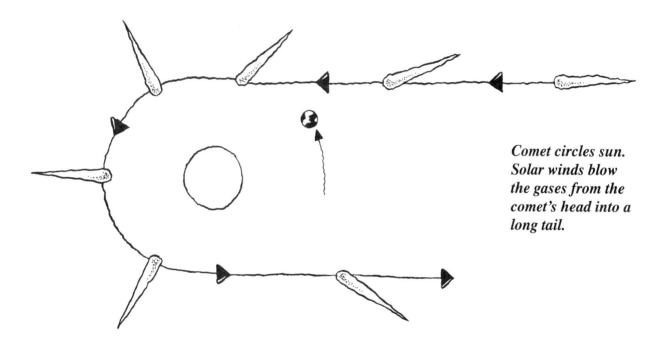

Comet circles sun. Solar winds blow the gases from the comet's head into a long tail.

word comet comes from the Greek word *kometes*, meaning "wearing long hair.") The tail always points away from the sun because the sun's *solar wind*, a strong stream of radiation, blows the gases backward off the coma.

A comet's tail can be 100 million miles long. Many comets have two tails—a straight one made of glowing gas and a curved one made of dust. The dust tail is usually yellowish-white because it shines by reflecting sunlight. A comet doesn't last forever. Each time it approaches the sun, it loses a part of itself. Eventually, it's reduced to a stream of fine dust and small rock bits.

Parts of a comet

HALLEY'S COMET

You've probably heard of Halley's comet. Although it's not the brightest or biggest comet, it's the most famous, because it can be seen easily with the naked eye. The first appearance of Halley's comet was recorded in 240 B.C. It was named after the seventeenth-century English astronomer Edmund Halley, who predicted it would return every 76 years. When Halley's comet visited in 1986, astronomers had their first close look at a comet—from the satellites that flew alongside it. Photographs showed the comet to be potato-shaped and made up primarily of ice.

Astronomer Edmund Halley discovers large comet.

HUNTING COMETS

Of the billions of comets in outer space, you'll only see one with a long, brilliant tail about once every ten years. Astronomers can predict when certain comets are due to appear in our skies. But a dozen or so smaller comets are discovered every year, usually by amateur astronomers. If you find a new one (you'll need binoculars or a telescope, and a lot of patience), it'll be named after you. Report your sighting to the nearest observatory or astronomy society.

To hunt for comets, scan the skies for a softly glowing object that doesn't appear on a night-sky atlas. It will change its position relative to the stars around it within a couple of hours. If you detect a tail, you've definitely found a comet. Comets may stay in your view for a couple of days or even weeks. The Great Comet of 1811 had a tail that stretched across almost the entire night sky and was visible for a year and a half!

Novae

A thrilling, but infrequent, discovery is a *nova*. Suddenly, a star will get very bright, increasing in magnitude about ten times. That's when it is given the "nova" name. Only binary stars (a pair of stars that orbit each other) can become novae and only when they're close to each other. Gas spills over from one star onto the surface of its companion star (usually a dwarf star) causing a violent nuclear explosion. The dwarf suddenly flares up, increasing in brilliance in the sky, making it seem as if a new star has been born (*nova* is Latin for new). Novae have been recorded in Hercules (1934), Puppis (1942), and Corona Borealis (1946). In 1975, a brilliant nova in the constellation Cygnus was bright enough to be seen with the naked eye. Some novae fade back to their original brightness in a few months or years, never to be noticed again. A few, however, periodically recur. It has been suggested by some astronomers that the Star of Bethlehem was either a nova or comet.

Star going nova. When gas from one star of a binary spills over onto its companion white dwarf star and causes it to suddenly explode, the flare-up of the star is called a nova.

The Lives of Stars

Stars come in all sizes. Some are smaller than the Earth, while others are larger than the Earth's orbit. Stars don't burn forever. Some stars last only a few million years, but some shine for hundreds of billions of years.

Right now, as you stare into space, stars are being born and old stars are dying. Although you can't see the process, you can find the birthplaces of stars in certain areas of the sky. The spaces between stars aren't empty. Rather, they are littered with clouds of gas and dust called *nebulae*. Nebulae comes from the Latin for "mist or cloud," and it's in some of these huge swirling clouds that stars begin their life.

NEBULAE

A star begins when gravity gradually pulls small clumps of gas and dust into a hot, dense ball, called a *protostar*. The ball becomes hotter and, eventually, nuclear reactions begin in its center. The explosions release energy as heat and light, and a star is born.

A star is born. A nebula, a swirling cloud of gas and dust (A), compacts (B) into a protostar (C), and a new star is formed (D).

Nebulae have some fascinating shapes and you can often see them with the naked eyes or with binoculars. If you search the area just below Orion's belt, you'll find the Orion Nebula. It looks like a fuzzy greenish patch. With binoculars, look for the Lagoon Nebula in Sagittarius, and in Cygnus for the North America Nebula, which is shaped like the North American continent.

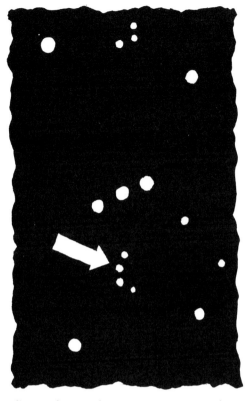

Orion and its Great Nebula

Horsehead Nebula—a dark nebula visible because of the bright nebulae behind it

Sometimes there are no stars in a nebula and the dark cloud can only be seen in silhouette, outlined against a lighter background. The most famous dark nebula is the Horsehead Nebula in Orion. It looks like a knight piece from a chess set.

Nebulae can gather around old, dying stars. As stars use up all their fuel, which is mostly hydrogen gas, they begin to expand. What happens next depends on the star's original size. A medium-sized star, like our sun, will swell up to form a red giant. Then it will shed its outer layers, while the center collapses, and become a

A star dies. A medium-sized star (A), like our sun, swells to a red giant (B), then shrinks to a white dwarf (C), and finally cools to become a black dwarf (D).

93

Ring Nebula. You'll need a telescope to see this planetary nebula in the constellation Lyra.

white dwarf star. A collapsing star is still very hot and continues to shine, like an ember in a fireplace, while it cools down. After billions of years, it will cool to a *black dwarf* star.

Meanwhile, the outer layers of gas are shed as a smoke-ring-shaped shell of gas that drifts off into space. Some of this gas may eventually condense to form new stars. Early observers thought these greenish-blue shells looked like the planet Uranus and called them *planetary* nebulae.

One such planetary nebula is the Dumbbell Nebula, which looks like a dumbbell-shaped, misty patch in the constellation *Vulpecula*, the Fox, just outside Cygnus. You'll need a small telescope to see the planetary Ring Nebula in Lyra, with its classic "smoke ring" shape.

SUPERNOVA TO BLACK HOLE

Giant stars die dramatically. They swell up into red supergiants, then explode violently, flinging most of their remains into far-distant space. When a star experiences this kind of explosion it is called a *supernova*. There's no mistaking a supernova when it happens because it shines brighter than 100 million stars! The best-known supernova appeared in 1054 A.D. It outshone everything in the night sky for two years.

Death of a giant star. Compared to a star the size of our sun (A), a giant star (B) swells to a red supergiant (C), and then explodes into a supernova (D).

A   B   C   D

The eruption of a supernova leaves behind an extremely dense core of gases and dust called a *neutron* star. A neutron star is tiny, millions of times smaller than a white dwarf star, which is about the size of the Earth. A neutron star is about the size of one of Earth's mountains.

Sometimes, a core collapses even more to form the most mysterious of all space objects, the *black hole*. The spinning gases of

94

Black hole swallowing gases from a passing star. The gravity of the tiny but very dense star at its center is so strong, not even its own light can escape.

this tiny star are so squashed by its strong gravity that nothing can escape, not even the light it makes. Since black holes don't shine, you're probably wondering how astronomers even know they exist. If any celestial object (like a companion star) gets near a black hole, it is pulled in. A sudden flare-up, followed by the disappearance of the light of the trapped star, is good evidence of a black hole. But astronomers still don't understand everything about the birth of stars and black holes.

What if you see a light in the night sky that doesn't fit anything we've talked about? Maybe you've found a UFO, or unidentified flying object. Considering how vast outer space is, and that Earth's space probes haven't yet searched beyond our own solar system, it's possible that extraterrestrial life does exist. But these aliens probably wouldn't look anything like the life found on Earth.

Case #8: Catch a Falling Star

From the chart below, mark the dates and shower names on a calendar so that you will know when to watch for "falling stars." You'll probably want to set an alarm to wake you up, because the peak of meteor shower activity is from 1:00 A.M. to dawn. Plan to watch for at least an hour, using just your eyes. In the Northern Hemisphere, look high in the sky to the east and southeast; in the Southern Hemisphere, look at the west and northwest sky. Some years, the show will be more dramatic than others, but even seeing a modest display will usually excite a sky detective.

IMPORTANT METEOR SHOWERS

NAME OF SHOWER	DATE OF MAXIMUM	NO. PER HOUR (APPROX.)	NO. YOU SEE PER HOUR
Quadrantids	Jan. 1–3	50	
Lyrids	Apr. 22	15	
Aquarids	May 4–6	35	
Perseids	Aug. 12	75	
Draconids	Oct. 9	100	
Orionids	Oct. 22	25	
Taurids	Nov. 5	10	
Leonids	Nov. 16	15	
Geminids	Dec. 13	75	

Case #9: False Evidence

How many mistakes can you find in this view of the night sky? Look for six false pieces of evidence.

Case #10: Aurora Stakeout

I You've just heard on the radio that massive sunspot activity has occurred and its effects should be felt on Earth within the next few nights. You've been contacted by the National Science Bureau to put your sky detective skills to work. Your job is to prove and to identify the presence of auroras. Begin your surveillance at sundown.

To make positive identification:

You'll need a camera, cable release, tripod, and color film to solve this case. For proof that you've witnessed an aurora, you'll need to catch these heavenly lights on film:

① Set the camera on a tripod or stable structure.

② Attach a cable release to the camera's shutter release button.

③ Open the camera lens to the largest f-stop (for example, f/2.8 or f/4.5).

④ Point the camera towards the brightest spot in the display.

⑤ Set the focus to ∞ infinity.

⑥ Expose the film at these suggested times. You'll have to experiment with the exposure times to be sure you catch these lights clearly on film. Try:
> Frame #1 at 1/15 second
> Frame #2 at 1 second
> Frame #3 at 5 seconds
> Frame #4 at 10 seconds
> Frame #5 at 20 seconds
> Frame #6 at 40 seconds

NOTE: Because auroras move, exposures longer than two minutes are not recommended.

II Good detectives make detailed notes of their observations while working on a case.

To record your observations:

If you see an aurora, the chart below might help you to positively identify the forms, colors, and intensities you see. A flashlight covered with a paper bag or a piece of red cellophane will help you jot down your notes in the dark while keeping your eyes "dark-adapted."

AURORA LOG

FORM	COLORS	DURATION	NOISE
GLOW			
ARCS			
RAYS			
BANDS			
VEIL			

Aurora forms (one may change into another):

Glow: A faint white light shining up from below the horizon.

Arcs: Rainbow-shaped lights that may flash up and disappear every 10 to 30 seconds.

Rays: Can look like searchlight beams and may shoot upward from an arc. Some may look like an opened fan.

Bands: Often resemble a huge curtain or drapery that "ripples" across the sky.

Veil: A cloudlike covering over large parts of the sky. Veils often appear after rays and are colorful.

•6•
THE TWILIGHT HUNT

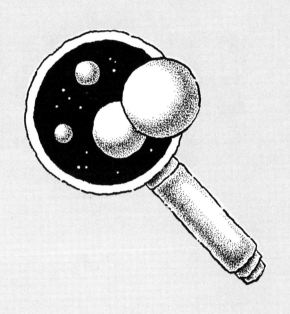

The sun doesn't shine for the Earth alone. There are eight other sunlit planets in our Solar System, all held captive by the sun's gravity. Many of the planets will frequently pass through your night sky during their travels around the sun. Martian canals, Earth's "twin," and the King of the Heavens are just some of the planetary sights that await your discovery. Also, take time out on your hunt to peer into the middle of your own Milky Way Galaxy, the hub of billions of stars.

Begin the hunt for planets at twilight, as the sky deepens to a dusky purple. About one-half hour after sunset, you may briefly glimpse the planet **Mercury**. Look for an orange-colored star that's as bright as Arcturus shining low on the horizon. You can also catch a glimpse of Mercury shortly before sunrise.

Look for Mercury low on the horizon, either right after sundown or just before sunrise.

Mercury

Mercury's moonlike surface

Mercury is a small, swift planet that orbits closest to the sun of all the planets in our sun's system. Our *Solar System* is made up of the sun, the nine planets and their moons, comets, meteoroids, asteroids, and interstellar dust. The planets all revolve around the sun in the same direction, from west to east, and get their names from Roman gods. The god Mercury was very quick, with wings on his heels, so served as messenger of the other gods.

The planet Mercury is an almost airless world with a surface temperature broiling at 800° F (430° C) on the day side, under the heat of the sun, and plunging to below −280° F (−175° C) in the shadow that is night. If you were to look at a photograph of Mercury's surface, you might mistake it for the moon. Like the moon's surface, Mercury is pitted with craters from debris that rained down during the formation of the Solar System.

WANDERING STARS

How can you tell a bright star from a planet? The best clue is to watch for a "star" that slowly and continuously changes position, or wanders, through various constellations. You'll probably need to keep track of these positions in a notebook over the course of a week or more to really detect the motion. Early astronomers noticed how these bodies moved and called them "planets," which in Latin means "wanderers." The planets, along with the sun and the moon, all move across the sky in a narrow zone known as the *ecliptic*. Over the year, this path takes the planets through twelve constellations called the *zodiac*. The band of stars that make up the zodiac lie close to the horizon and include such constellations as Libra, Gemini, Pisces, and Aries.

Solar System. In order from sun: Mercury, Venus, Earth, Mars, Jupiter, Saturn, Uranus, and Neptune and little Pluto (which sometimes switch places).

FORETELLING THE FUTURE

The rulers of the ancient Greek, Mayan, Chinese, and Egyptian cultures believed that the motions of the sun, moon, and planets directly controlled events on Earth. The beasts and figures they imagined in the stars of the zodiac were responsible for the sun's safe passage through the sky. *Astrologers*, people who chart these movements to foretell the future, were often among the most

important members of the royal court. But it was sometimes a dangerous job—emperors were known to execute an astrologer for failing to predict a catastrophe, such as a flood or drought.

Many people today still believe their destiny is influenced by planetary positions and consult astrologers regularly. Others just read their astrology prediction, or *horoscope*, in the local newspaper.

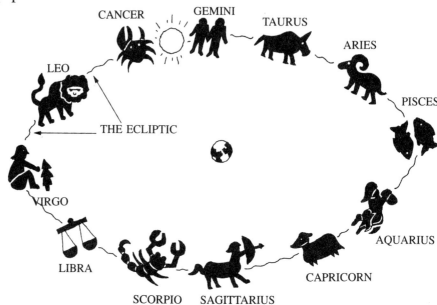

Zodiac. A band of twelve constellations through which the sun, moon, and planets appear to move.

Earth's Sister Planet

It's easy to find the planet **Venus** in the twilight sky. Named for the goddess of beauty, it surpasses every other star with its silvery brilliance. Venus is most dramatic when it falls close to a waxing crescent moon. This planet shines so brightly because it's wrapped in a thick, heavy blanket of white clouds that strongly reflect sunlight, the way fresh snow does. Like the moon, Venus goes through phases as it travels around the sun, but you'll need a good pair of binoculars and steady hands to see those phases.

Venus and the crescent moon

INFERIOR AND SUPERIOR

Mercury and Venus orbit closer to the sun than Earth does and are called the *inferior planets*. There are certain years when a particular planet will pass closer to the Earth, and, therefore, seem brighter. The sun blocks our view of the inferior planets when they are closest to Earth. When they are in the best viewing position

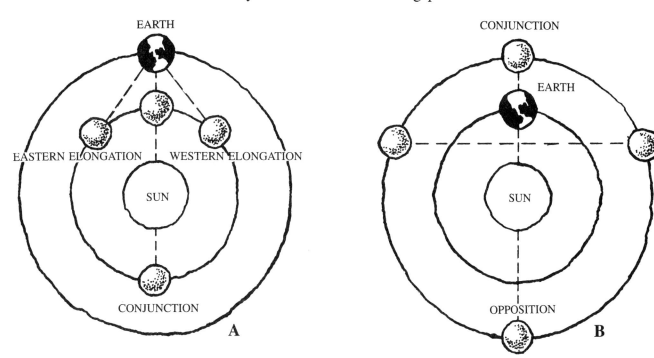

from Earth, they are in what astronomers call the *greatest elongation* positions. The planets that orbit beyond the Earth are called the *superior planets*. The best time to see one of the superior planets is when it is in position opposite the sun, or in *opposition*.

THE SURFACE OF VENUS

What type of world was hidden beneath Venus' clouds remained a mystery until the space shuttle launched the Magellan spacecraft on May 4, 1989. This probe bombarded Venus with thousands of radio signals per second to form detailed pictures of its entire surface. And the results are exciting. In some ways, Venus could be Earth's twin. It's about the same size and density as the Earth, and its orbit is closest to Earth's. Venus is a rocky world covered with craters, rivers of lava, mountains, and valleys—all very similar to geological features on Earth. However, Venus has no water. Its

Locating the planets. Whether or not you can see a planet in the night sky depends on where it lies along its orbit. Inferior planets (A), closer to the sun, are best seen at greatest elongation positions; superior planets (B), farther from the sun, are best seen when they are in opposition.

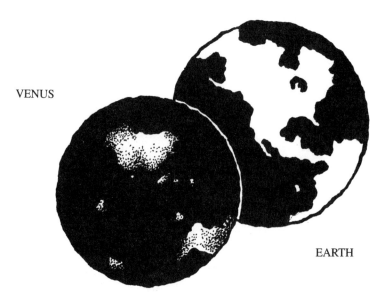

Earth and her "twin," Venus

VENUS

EARTH

surface temperature, 900° F (468° C), is hot enough to melt lead, and it has an atmosphere that is 95 percent carbon dioxide (Earth's atmosphere is mostly nitrogen). The conditions now on Venus might be what could happen on Earth if there was a runaway greenhouse effect. Scientists hope to learn more about Earth's geology and climate by studying the Magellan Mission to Venus results in more detail.

Venus orbits the sun in 225 days. Because it is the slowest-rotating body in the Solar System (one day on Venus takes 243 Earth days), it is actually spinning slowly *backwards*.

Are you an early riser? If so, look for Venus shining brightly in the dawn sky, when it is known as the Morning Star. Once upon a time, the Pawnee Indians would honor the Morning Star with an important ceremony. An intricate ritual was performed every few years, which included the killing of a maiden from a neighboring tribe so that her soul would be released and rise up to the Morning Star. This sacrifice was supposed to ensure the fertility of the land and an abundance of buffalo so that the tribe would survive.

Pawnee Morning Star ceremony

The Red Planet Mars

Can you spot the fiery red planet **Mars**? You might mistake it for the star Antares (which means "rival of Mars"), but its changing position amongst the stars is the giveaway that you've found a planet. Mars reminded Roman star watchers of blood, so they named it after their war god. Mars gets its reddish appearance from the dust storms that sweep up clouds of the iron-rich dirt that coats the planet's surface.

Mars. Don't be fooled into thinking it's a red star. Go on surveillance. A constant movement through the zodiac will reveal the suspect "star" to be the planet Mars.

THE MARTIAN SURFACE

"There's life on Mars," blared newspaper headlines in 1886. The evidence was a network of "canals" spotted through early telescopes. Some people thought these "canals" must have been built by Martians to transport water from the polar ice caps of that planet. But the Mariner and Viking probes of 1965 and 1976 have proven that no life exists on Mars.

Mars is a parched, rust-covered world littered with boulders, and riddled with craters. These craters are less jagged than those found on Mercury or on the Moon. Because Mars has some atmosphere, its winds erode and smooth the crater surfaces. Another interesting Martian feature is the polar caps, made up of frozen water and carbon dioxide (dry ice), which melt and refreeze seasonally.

Mars' surface. The polar caps wax and wane with the seasons: they're large in winter and shrink during the summer.

If you were to land on Mars, you'd see the biggest known mountain in the Solar System. The *Olympus Mon* is a huge ancient volcano. Its base is about the size of Ireland. It rises a spectacular 100,000 feet (30.5 km) high. Our famed Mount Everest rises only to 29,000 feet (8.8 km). Another awesome sight is the Mariner Valley, which is five times deeper than the Grand Canyon.

Olympus Mon, a Martian volcano

THE TERRESTRIALS

Mercury, Venus, Earth, and Mars are iron-rich, rocky bodies. They're called the *terrestrial* planets because they consist of materials similar to those found on Earth. The Solar System formed about 4.6 billion years ago. The solar nebula, a disk-shaped cloud of gases and dust, surrounded the sun. Over time, the particles in

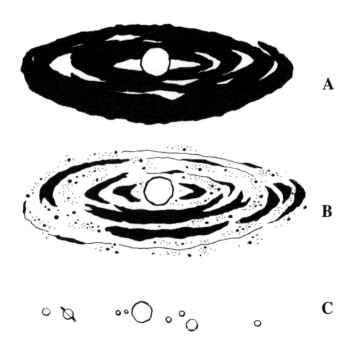

A

B

Birth of the Solar System. A dense nebula circles the sun (A), the nebula condenses into larger particles (B), and the planets and their moons are formed (C).

C

108

the nebula condensed and were pulled together by gravity to form the individual planets. It's believed that at one time all the planets were made up of a rocky core surrounded by thick layers of gases. The planets closest to the sun, the terrestrial planets, gradually lost their covering of thick gases, which were blown away by the solar wind.

The Asteroids

Beyond Mars is a belt of irregular chunks of rock called *asteroids*, which means "starlike." There are hundreds of thousands of these asteroids circling the sun, and most are only a few kilometres wide. All the asteroids put together wouldn't have enough mass to form an object the size of Earth's moon. Their origins still puzzle

Asteroid belt. The asteroids orbit the sun in a belt between Mars and Jupiter.

astronomers, but they're probably pieces of a larger body or bodies that broke up as the planets were forming.

You'll need to check astronomy periodicals to know where in the night sky to look for the largest asteroids. You can't see them with your naked eyes, and through binoculars they look just like faint stars. *Ceres*, named for the goddess of agriculture, is the largest asteroid, with a diameter of 600 miles (1000 km), about the width of Texas.

Asteroids have crossed Earth's path in the past, and some have fallen to Earth as meteorites. Many scientists believe one of these impacts caused the extinction of the dinosaurs, some 65 million

Asteroid falling into ocean. The dinosaurs may have been "killed off" by a huge asteroid crashing to Earth.

years ago. One theory is that the asteroid-meteorite crashed into the oceans, sending up thick clouds of steam and smoke that blocked out most of the sunlight for several years. Without sunlight, the plants died. So both the dinosaurs that ate the plants and the meat-eating dinosaurs that lived off the plant-eating animals died out.

Jupiter

Past the asteroid belt are the *Jovian* planets: Jupiter, Saturn, Uranus, and Neptune. These gaseous giants are cold bodies that have rings and many moons orbiting them.

It's easy to locate the King of the Heavens, **Jupiter**, as it wanders through the zodiac. If you find a brilliant yellowish-white star in the heavens, you're looking at the largest body in the Solar System. Jupiter's diameter is eleven times that of Earth's, about

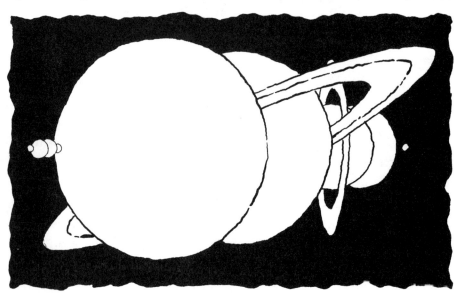

Jovian planets. Compare their sizes with those of the four terrestrial planets at the left side of the drawing.

the size of a basketball compared to a golf ball. In 1610, Galileo first identified Jupiter and four of its moons, *Io*, *Callisto*, *Ganymede*, and *Europa*. With very steady binoculars you can see these four moons strung out in a line on either side of the planet. They appear as tiny pinpricks of light that shift positions every hour. Jupiter has at least sixteen moons, and more will probably be found.

Jupiter and four moons

Jupiter is not solid like Earth. Rather, it is a giant ball of liquid hydrogen, with a small rocky core and blanketed with ivory- and salmon-colored clouds. The planet rotates rapidly. Its day is less than ten hours long. This rapid spinning causes Jupiter to bulge along its equator and produces powerful winds that swirl the clouds into ever-changing bands and streaks. A fascinating feature of Jupiter is the *Great Red Spot*, which looks like a red, beady eye on its Southern Hemisphere. It's thought to be a huge spiralling storm cloud.

Surface of Jupiter, showing Great Red Spot

Voyager spacecraft. The Voyager I and II spacecraft have been the most successful planetary probes launched from Earth.

Have you seen the dramatic photographs sent back to Earth from the Voyager I and II spacecraft flybys? The Voyager missions were launched in 1976 and 1977 to gather data on the outer planets. The details and clarity of the photographs sent back to Earth continue to awe everyone. While photographing Jupiter from its dark side, Voyager I discovered the presence of a thin ring around the planet.

Saturn

Saturn

Another whitish star you can see with your naked eyes is the second-largest planet, **Saturn**. You're probably familiar with this planet because of the many pictures of its exotic system of rings. Binoculars show Saturn only as a flattened disk, so if you have the chance to look through a telescope to see the rings, don't pass it up.

Saturn is a cloud-covered ball of liquid hydrogen, like Jupiter, but it is a quieter and less turbulent world. It is named for the Roman god of farming. Saturn is very far from the sun, so it is very cold (−300° F, −185° C).

The Voyager spacecrafts unveiled a goldmine of information about Saturn's rings. What we previously thought were three rings are actually hundreds of rings that look much like the grooves in a phonograph record. The rings aren't solid but are composed of trillions of ice particles. This spectacular ring system is quite wide, stretching 94,000 miles (150,000 km), but it is only a mile or less (1–2 km) thick!

Close-up of Saturn's rings. Probes have revealed that Saturn's two rings, visible through telescopes, are really hundreds of rings made up of trillions of ice particles.

Saturn has at least 20 moons, more moons than any other planet. One moon, *Titan*, is especially interesting to astronomers because it has an atmosphere that is 95 percent nitrogen gas, somewhat like Earth's atmosphere, and could possibly support life. One goal for future space exploration is to probe extensively beneath the thick nitrogen blanket that surrounds Titan.

DANCING PLANETS

If you watch the movements of the superior planets over a period of a few months, you may be surprised to see them sometimes dance or move backwards in the sky. Instead of continuing their journey straight across the sky from west to east, they seem to slow down, stop, and then move backwards—going from east to west. Then they stop and begin to move again in the original, west-to-east direction. This "backwards" dance is called *retrograde motion*, and is only an optical illusion.

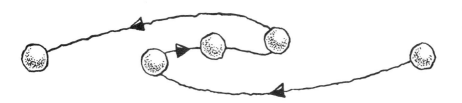

Retrograde motion. Planets sometimes seem to move backwards in the sky.

The farther a planet is from the sun, the longer it takes to travel around it. Jupiter orbits the sun once every 11.9 Earth years, and Saturn takes 29.5 Earth years for one orbit. Retrograde motion happens when Earth overtakes these planets in their travels. For instance, when Earth because of its faster speed in orbit catches up with Jupiter, this giant planet will appear to slow down and move backwards to anyone watching from Earth. As Earth moves farther along in its orbit, Jupiter will appear to resume its usual path.

Uranus and Neptune

It's very difficult to distinguish the planets **Uranus** and **Neptune** from the other faint stars in the heavens. The Voyager II has added some information about these little-known planets from the flybys in the late 1980s. We already knew that it takes Uranus, called "the god of heaven," 84 years to circle the sun. Voyager photographs show this gassy, aquamarine planet to have a thick, soupy atmosphere of methane gas. Uranus lies on its side. Its axis is tipped just eight degrees from being horizontal. Some believe that the planet got knocked on its side long ago during a collision with another object in space. There are at least fifteen moons and a system of ten skinny rings circling this cold planet.

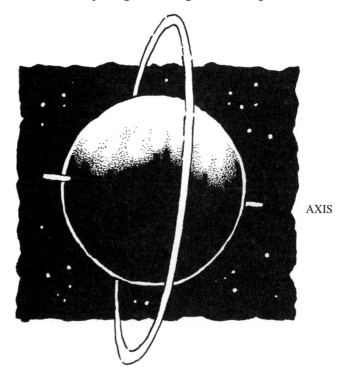

AXIS

Uranus and axial tilt. Unlike all the other planets in the solar system, the axis of Uranus is so tilted the planet is almost lying on its side.

Looking through a telescope like a blue-green speck, Neptune is almost a twin of Uranus. Neptune is named for the god of the sea. Along with its eight moons and an extremely thin system of rings, Neptune completes one trip around the sun every 165 years. Voyager photographs show the gaseous surface clouds of Neptune as being more banded than those of Uranus.

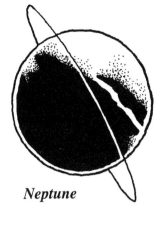
Neptune

Pluto

Far out in the Solar System is the planet **Pluto**, circling some 3.6 million miles (5.9 million km) from the sun. Pluto was discovered in 1930, when it was spotted passing through the constellation Gemini. Named after the god of the underworld, Pluto is a frigid, tiny planet, smaller than Earth's moon. It is similar to the terrestrial planets with its rocky makeup. It takes Pluto 248 years to revolve once around the sun. Because its orbit is off-centered, or *eccentric*, every now and then it sweeps inside the orbit of Neptune.

In 1987, Pluto's satellite, *Charon*, was discovered. Scientists aren't sure whether Charon and Pluto are actually a double planet system or whether some larger object speeding through space knocked a piece off of Pluto, which formed Charon. To date, no probes have passed Pluto and much remains to be learned about this very distant world.

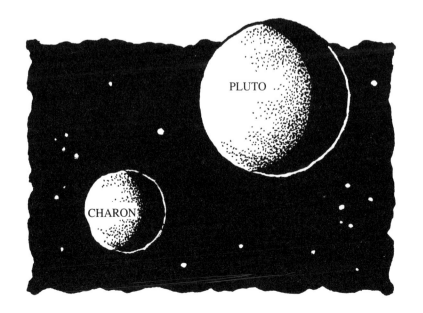

Pluto and its moon, Charon. Astronomers are still debating whether or not Pluto is truly a planet, and whether or not its moon, Charon, is really a broken-off piece of Pluto.

The Milky Way

Galaxies

While hunting for planets, you may detect an area of the sky that is so packed with stars that it looks hazy or milky. This broad river of light is a portion of the galaxy that Earth belongs to, the *Milky Way Galaxy*.

THE STARS AROUND US

A *galaxy* is a huge system of nebulae, star clusters, novae, binaries, and billions of stars. The Milky Way Galaxy is gigantic. It contains our entire Solar System plus all the stars you can see in the night sky. There are at least 200 billion stars whirling about in the Milky Way Galaxy.

Raven in snowshoes, an Eskimo legend

Some cultures of the past thought the Milky Way was a river or a path that the souls of the dead travelled along to reach the next world. The Plains Indians believed this cloudy part of the sky was the dust kicked up by a buffalo and a wildcat racing across the sky. The Eskimos believed it was the tracks made by the Raven's snowshoes as he walked across the sky while creating the inhabitants of the Earth. And natives of Patagonia, an area in southern Argentina and southern Chile, say these stars are the souls of the dead who roam and hunt for ostriches along this heavenly road.

Can you locate the zodiacal constellation of Sagittarius? Look for it on the southern horizon in late summer. This star-rich area is near the center of your galaxy. The Milky Way Galaxy is shaped like a huge spiral, with a thick, swirling soup of stars at the center. More stars and dust extend outward, along spiral arms that rotate around the nucleus. Our Solar System is near the inner edge of one of these arms. If you were in a rocket ship far out in space and could peer down on the Milky Way, it would look like a giant pinwheel. Seen from its side, it would look like a band of light with a bulge in the middle.

LOCATION OF THE SUN

Our Milky Way Galaxy viewed from above (top) *and from the side* (bottom)

FARAWAY PLACES

The Milky Way Galaxy is not alone in space. Astronomers believe there are billions and billions of galaxies. The largest galaxy to date was discovered in November 1990. *Abell 2029* is believed to be sixty times larger than the Milky Way and is a whirlpool of more than one trillion stars. Not all galaxies are shaped like the Milky Way. For American astronomer Edwin Hubble, the study of the shapes of galaxies and how galaxies came to be was his lifelong pursuit. He analyzed the light coming from galaxies to prove that the galaxies were all rushing away from one another. He took many photographs of galaxies, while working at the Mount Wilson Observatory in California, and identified three main types of galaxies in 1925:

The billions of galaxies in the universe come in many different shapes.

spirals: with a nucleus and spiralling arms
barred spirals: with spiralling arms extending from a thick central band of stars
elliptical: immense clouds of gas that look like a squashed ball

Elliptical galaxies are the most common and include some of the largest known galaxies. Two other minor categories of galaxies have been added since 1925:

irregular: with no definite shape
dwarf: small and dim

In 1990, the United States launched the most sophisticated satellite to date into space, the Hubble Space Telescope. Named in honor of Edwin Hubble, this telescope should reveal exciting discoveries as it peers into distant galaxies.

SPACE CLOUDS

If you live in the Southern Hemisphere, you can easily glimpse with the naked eye two nearby irregular galaxies, the *Large* and *Small Magellanic Clouds*. They were named after the Portuguese explorer Ferdinand Magellan, who observed them while sailing the southern seas in 1510. He called them clouds because of their soft outlines and glowing appearances. Look for the Large Magellanic Cloud in the southern constellations *Mensa*, the Table Mountain, and *Doradus*, the Swordfish. The Small Magellanic Cloud is found near *Tucana*, the Toucan. Both Magellanic Clouds lie within the gravitational field of the Milky Way Galaxy and, even though they are really galaxies, they orbit it as satellites.

Clouds of Magellan. The Large and Small Megellanic Clouds, two nearby galaxies, are a familiar sight to Southern Hemisphere sky detectives.

SMALL CLOUD

LARGE CLOUD

ANDROMEDA

To spy on the most distant object visible to the unaided eyes, look for the *Andromeda Galaxy*. During the autumn in the Northern Hemisphere, use the constellation *Pegasus* to guide you to constellation *Andromeda*, which extends from the upper left corner of Pegasus. The Andromeda Galaxy, a spiral galaxy, is within the constellation Andromeda. It forms a hazy patch up and to the right of the bright star Mirach, one of a triangle of stars. The light you see coming from the Andromeda Galaxy first started its journey more than two million years ago, during the time of the prehistoric man *Homo habilis*.

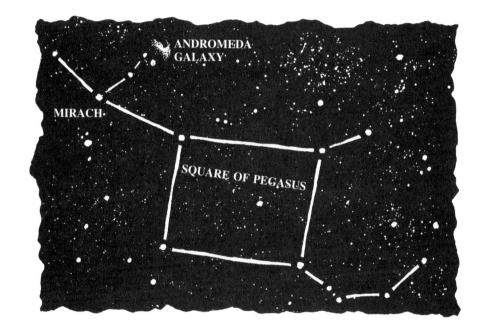

Locating the Andromeda Galaxy from Pegasus, up and to the right of Mirach in Andromeda

The Universe

Galaxies aren't scattered randomly throughout the universe. They journey through space in clusters. The Magellanic Clouds, the Andromeda Galaxy, the Triangulum Galaxy, and our own Milky Way Galaxy are part of about 30 galaxies known as the *Local Group*. Like other galaxy clusters, the Local Group is bound together by the forces of gravity. Scientists estimate that the *Universe*, which is literally everything in space, consists of 100 billion major galaxies.

STEADY STATE

Cosmologists, scientists who study how the universe as a whole came to be, have some clues and some guesses about the workings of the universe. Some of them believe that the universe has always existed in a state that is similar to what we see today. This *Steady State Theory* maintains that there was no beginning, nor will there be an end to the universe, and that the universe is expanding because matter is continually being created.

Local Group galaxies, travelling through space bound together by gravity

BIG BANG

Today, most cosmologists believe in the *Big Bang Theory*. They claim that all the matter in the universe was crammed together when it was formed, around 18 billion years ago. A massive explosion, called the Big Bang, caused the original mass to fly apart. Matter was spread throughout space as a thin gas that began

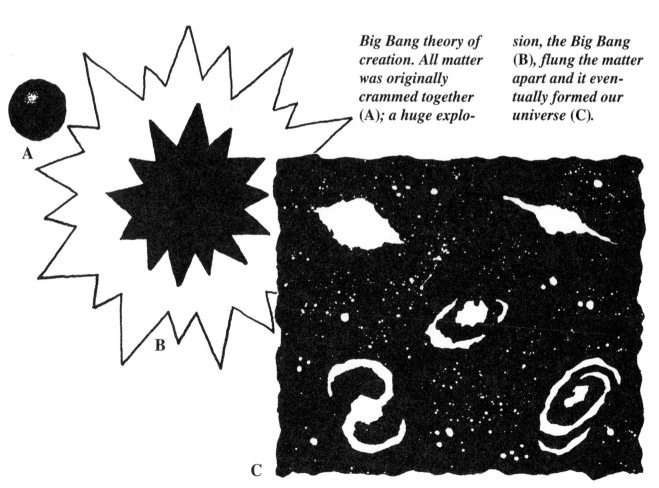

Big Bang theory of creation. All matter was originally crammed together (A); a huge explo-sion, the Big Bang (B), flung the matter apart and it even-tually formed our universe (C).

to cool and break up into clouds. About one billion years after the explosion, the clouds condensed into galaxies. Cosmologists are still trying to unravel the details of this process. By detecting the motion of objects in space, by using light, radio, and x-ray emissions, we do know that the clusters of galaxies are rushing apart from one another, and that the universe is expanding.

Local Group as part of the Universe

QUASARS

In 1963, radio astronomers detected starlike objects giving off fantastic amounts of energy. They named them quasi-stellar-sources, or *quasars*. Quasars shine as brightly as a trillion suns and are the most distant objects we can see. They are estimated to be 15 billion light-years away. This means that the light we see coming from the quasars left them when the universe was first forming! These blazes of light are probably from the energy thrown off by young galaxies as they formed after the Big Bang explosion.

Each year, new clues are revealed that help solve portions of this gigantic puzzle. But many, many mysteries about stars, planets, galaxies, and the universe remain unsolved.

Milky Way Galaxy as part of the Local Group of Galaxies

Your Place in the Universe

The universe is so vast that it's hard to grasp its size. The next time you're gazing at the night sky, see if you can imagine where you are on Earth relative to the rest of the universe.

Solar System as part of the Milky Way Galaxy

Earth as part of the Solar System

You on Planet Earth

Case #11: Along the Milky Way

Take a journey with your eyes along the Milky Way to view enormous starfields, nebulae, and clusters in this galaxy. To see the sights best, head out on a clear night and bring along a pair of binoculars. How much of the Milky Way you see depends on how far north or south you live. August is the best time for your "trip." This travel log highlights some of the sights, starting with the northern constellation of Cassiopeia and sweeping down through the southern constellation of Puppis.

GALAXY TRAVEL LOG

① The Milky Way flows down from the constellation Cassiopeia as a narrow trail of numerous, distinctly bright stars.

② Near Deneb, the top star in the constellation Cygnus (or Northern Cross), look for a dark nebula, the *Northern Coal Sack* (illus. **A**).

③ Through Cygnus, Sagitta, Aquila, Scutum, and Ophiuchus, the Milky Way seems to divide into two lengthwise sections. This cut is really only a lane of dark clouds and dust that blocks out starlight. It is known as the *Great Rift*.

④ Check Cygni, the bottom star of Cygnus, with binoculars for a double star. The main star is yellow with a 3.0 magnitude, and the companion star is bright blue with a 5.0 magnitude.

⑤ The western arm of the Milky Way fades and disappears near the constellation Serpens. The eastern arm thickens with brilliant starfields that increase in number farther south of the constellation.

⑥ The "milky" or hazy appearance is obvious around the constellation Sagittarius. The center of the Milky Way Galaxy lies in this bright region of the night sky. Sagittarius has more globular clusters than any other constellation. You won't need

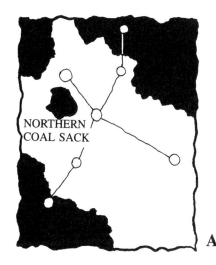
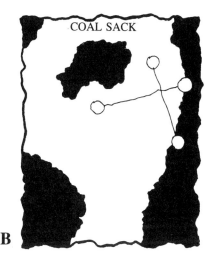

Along the Milky Way. Northern Coal Sack and Cygnus (left) *and Coal Sack and Southern Cross* (right)

binoculars to see the Lagoon Nebula, near the Archer's bow. However, binoculars will show the dark lane, or "lagoon," that slices this hazy, glowing cloud in half.

⑦ The Milky Way becomes broadest as it extends into the constellation Scorpius. The bright-red star Antares is surrounded by many clusters. Two spectacular open clusters, M6 and M7, can be found near the Scorpion's tail.

⑧ Southern sky detectives will notice that the Milky Way grows brighter by the constellation Crux. Look east of Crux for an area that appears completely dark to the naked eye. This dark nebula is known as the *Coal Sack* (illus. **B**). While you're sighting on Crux, be sure to look at the naked-eye cluster there called the Jewel Box.

⑨ Sweep down and through Crux to the constellations Carina, Vela, and Puppis for some of the most brilliant starfields. Look near the eastern edge of Puppis for a naked-eye cluster with a bright-orange supergiant star. Where Carina borders on Centaurus, try to find a very bright open cluster, called NGC 3532. This cluster has at least 400 stars, 60 of which you can easily see through binoculars.

A good sky atlas can help you find the exact locations of clusters and nebulae. You can usually find this reference book in your library. Also, check monthly astronomy magazines for detailed maps of the sky.

Case #12: Alien Encounter

I Astronomers have discovered what they believe to be a planet near the star Alpha Centauri. Satellite probes have gathered some information about it. Your detective agency has been chosen to investigate the possibility of life on this mystery planet. As leader of this space mission, you must decide what items should be brought along aboard the spaceship. Based on these five clues, choose your planet-exploring equipment from the list below.

Clues:
- has the gravity of Earth's moon
- has an atmosphere that's similar to Saturn
- has the same temperature found on Venus
- has a surface like Jupiter
- revolves around its sun but doesn't rotate

Equipment checklist:

① climbing rope and hiking boots

② oxygen tanks

③ sunglasses

④ dune buggy

⑤ weight belt

⑥ heaters

⑦ boat and diving gear

⑧ asbestos spacesuit

⑨ flashlight

II What do you imagine an alien from this planet would look like? Make a sketch showing some of the features that would help it survive in its environment.

A universe made for sky detectives

INDEX